Springer Biographies

The books published in the Springer Biographies tell of the life and work of scholars, innovators, and pioneers in all fields of learning and throughout the ages. Prominent scientists and philosophers will feature, but so too will lesser known personalities whose significant contributions deserve greater recognition and whose remarkable life stories will stir and motivate readers. Authored by historians and other academic writers, the volumes describe and analyse the main achievements of their subjects in manner accessible to nonspecialists, interweaving these with salient aspects of the protagonists' personal lives. Autobiographies and memoirs also fall into the scope of the series.

Herwig Schopper · James Gillies

Herwig Schopper

Scientist and Diplomat in a Changing World

 Springer

Herwig Schopper
CERN European Organization for Nuclear
Research
Geneva, Switzerland

James Gillies
CERN European Organization for Nuclear
Research
Geneva, Switzerland

After the initial publication of this book, few corrections which don't change the meaning have been made, typos have been corrected.

ISSN 2365-0613   ISSN 2365-0621 (electronic)
Springer Biographies
ISBN 978-3-031-51044-1   ISBN 978-3-031-51042-7 (eBook)
https://doi.org/10.1007/978-3-031-51042-7

This Springer imprint is published by the registered company Springer Nature Switzerland AG
The registered company address is: Gewerbestrasse 11, 6330 Cham, Switzerland

Paper in this product is recyclable.

Foreword

When I first came to CERN in 1994, Herwig Schopper had already retired—although for Herwig retirement simply meant embarking on a new career that drew on his strengths and experience as a scientist and as a manager of science. It is through this post-retirement career as a pioneering science diplomat that I have come to know and appreciate Herwig both as an extremely valuable and knowledgeable colleague and as a kind and generous human being.

Through his career, Herwig has played a leading role in national institutions in Germany, as well as at international organisations including DESY, SESAME (Synchrotron-light for Experimental Science and Applications in the Middle East) and CERN, where he served as Director-General from 1981 to 1988.

Although Herwig's time at CERN represents but one aspect of his long career, it is the one with which I am most familiar. First coming to CERN in the 1960s, Herwig built strong relationships between the Laboratory and his home institutes in Germany. When he was appointed Director-General in 1980, he took the helm at a pivotal moment for CERN. Tasked not only with building the Large Electron Positron collider, LEP, the Lab's most ambitious project until that time, he also had the job of bringing the two CERN laboratories in Meyrin and Prévessin together under a single management structure. To say that his mandate was a success would be a great understatement. With the completion of LEP, which started up in 1989, he achieved a delicate balancing act between the scientific ambition of the physics community and the economic requirements of the Member States.

All this is a matter of record, and I would like to end this brief foreword on a more personal note. At CERN, I have had the honour to serve as Spokesperson of the ATLAS Collaboration and Director-General of the Laboratory. These are challenging roles, and Herwig has always been there to share his thoughts and experience. He has great vision, and there have been many occasions on which I have been grateful for his advice and support. He has also been a demanding visitor to my office on occasion, petitioning for CERN to support his more recent endeavours to deploy science in the service of peace. The SESAME Laboratory is a case in point. It was a pleasure for me to represent CERN at the inauguration of SESAME in 2017 and to see the Laboratory thriving today. I wish Herwig similar success with his current projects and invite you to discover a glimpse of his extraordinary life through the pages of this book.

Fabiola Gianotti
Director-General, CERN
Geneva, Switzerland

Authors' Note

History is that certainty produced at the point where the imperfections of memory meet the inadequacies of documentation.

—*Julian Barnes*

Notes from James Gillies

In 1986, the first time I came to CERN, Herwig Schopper was the Director-General. At the time, the idea that I might one day ascend to the dizzying heights of the 5th floor of CERN's main building, let alone meet the DG and one day help him to set out his memoir would have seemed laughable. It has been an immense privilege and a pleasure to do so. The structure of this book is Herwig's wish. Most of the writing was done during the pandemic years, so we had long conversations on Zoom, which were then transcribed and written up as chapters, each with an anecdotal 'in his own words' section. We iterated each chapter until Herwig was happy with it, and when eleven chapters were complete, Herwig contributed the last, consisting of reflections on his life, and thoughts he would like to share with younger generations. As he says himself, younger generations tend to see the past as a collection of quaint stories with little relevance to them. If I've learned anything from this project, we all have much to learn from those who have lived, in particular, those who have lived as fully as Herwig Schopper.

Among those who helped us are the librarians and archivists at CERN and UNESCO, in particular Sandrine Reyes and Jens Vigen at CERN, and Eng Sengsavang at UNESCO. The Cavendish Laboratory, DESY, KIT, KTH, the Imperial War Museum, the Potsdam Museum, the Harwell and Rutherford Appleton Laboratories, the Městské Muzeum, Lanškroun, the Max Planck Institute for Plasma Physics, the University of Mainz, the Fuldaer Zeitung, the Swiss Federal Department of Foreign Affairs, the APS News staff, Padro Abreu at LIP and the Embassy of Chile in Geneva all helped us with our searches. While not all searches were successful, the assistance we received is very much appreciated. CERN's graphic design office helped with some of the illustrations, and our thanks go to Fabienne Landua and Alice Duc for this.

Another person we'd like to thank is Alan Watson, who not only helped piece together the story about a meeting between Jim Cronin and Carlos Menem but also shared Julian Barnes's most apt quote about the nature of history.

This is a phenomenon that the authors encountered on many occasions in the writing of this book—particularly since a recurring theme of Herwig Schopper's narrative is that much of what is important happens off the archived record—discussions in the margins of meetings, or agreements made over coffee. As a consequence, many of the stories he tells in this book required considerable detective work to verify the details. We are grateful to the many people who helped us piece together elements of the written record in support of Herwig's recollections. While we do not claim that the result represents absolute historical certainty on all counts, we are at least confident that it is a faithful record of Herwig's long and eventful life.

Last but not least, we would like to thank CERN for covering the open access fee for this book, making it accessible to potential readers across the world.

Notes from Herwig Schopper

This biography would not have been written without the exceptional collaboration with James Gillies. Unfortunately, CERN is too big for the Director-General to have personal contacts with all active physicists. So I got to know James only when he had changed from doing physics to public relations and I had become involved in the SESAME project. Three of the founding fathers of SESAME considered writing the history of the project with the help of James. This idea is on hold, but it had the result that I got to know James quite well. Some years later, at a lunch with Christian Caron from Springer, Jens Vigen from CERN and a few other colleagues, the suggestion was made to me that I should write my autobiography. I was very reluctant and said I could only consider it if I could have substantial support from somebody who has a solid knowledge of physics and at the same time experience in public relations. After a relatively short discussions James Gillies's name was proposed. Knowing him from SESAME, I immediately agreed. After lunch I contacted James and to my great relief he accepted, and thus an extremely fertile and from a personal point of view, pleasant, cooperation started. At first I dictated some chapters and we had them transcribed, but as time went on, the process evolved to a long series of interviews and subsequent discussions to produce the texts of the various chapters. To my great satisfaction, James immediately understood my intentions and very often expressed them much more clearly than I ever could have done. As we came to know each other better, James's editorial eye started to go further than the purely linguistic, and we'd discuss the best way to express what I wanted to say. He also complemented the text in many respects by researching other sources, thereby helping me to fill the gaps in my memory. As a result, James is now perhaps the person who knows my life better than anyone, including private and professional facts. We thus became through the cooperation friends. Covering my hundred years he also became an expert for the development of science and particularly physics. Including the history of national and international laboratories, he is now an expert in this field and I hope that this experience will be used in future for other historical works.

Contents

Chapter 1
Early Years and Private Life

February 28, 1924 was a winter's day much like any other in the Czechoslovakian town of Landskron. The world was at peace, if somewhat unstably, and as a sign of global inclusiveness, a celebration of sport, the first Winter Olympic games, had just concluded in Chamonix. Norway's 14 athletes topped the table with a 17-medal haul, although hardly anyone in Landskron noticed—news travelled more slowly in those days. It was also the day on which Franz and Margarethe Schopper welcomed their son Herwig into the world.

In 1924, Europe was still reeling from the Great War, which had seen the old order swept away. Gone was Austria-Hungary, which had dominated central European politics since 1867. Gone was tsarist Russia, replaced by the USSR. Britain's imperial power would never recover, and a chastened Germany was suffering from the punitive Treaty of Versailles. Largely ostracised by the victors, Germany was barred from competing in those first winter games, although several of the countries that emerged from the ashes of Austria-Hungary were there, including Czechoslovakia.

Austria-Hungary had given way to nation states, with the peoples of the former empire scattered throughout. That's how Viennese teacher, Franz Schopper, found himself teaching at the Landskron secondary school, or gymnasium, after the war. In Landskron, returning as a prisoner of war from Russia, he met Margarethe Stark, a local woman ten years his junior, and it was not long before they were married.

Landskron was a German-speaking town in the Czechoslovakian region of Bohemia, close to Moravia, and to the Polish and German borders. As Herwig recalled from his youth, information about the world beyond those borders was sparse: "There was no television, there were no mobile phones, very few telephones at all, and practically the only information we got was the local newspaper. Radio was practically unknown and it was very bad. There were some people who could listen to the BBC, but they were the exception. I lived just a few kilometres from the German border but to go to Germany at that time was completely impossible."

© The Author(s) 2024
H. Schopper and J. Gillies, *Herwig Schopper*, Springer Biographies,
https://doi.org/10.1007/978-3-031-51042-7_1

Fig. 1.1 Herwig's father Franz in his library at Annahof, taken around 1970 (Herwig Schopper's personal collection. ©Herwig Schopper, All rights reserved)

Czechoslovakia's borders had been carved out by the Great War's victors who had made Tomáš Masaryk its interim leader. The fate of the fledgling nation was very much in flux, and this would come to have a profound influence on the young Herwig as he grew into adulthood. In 1924, big changes were still well over a decade into the future, however, and Herwig was blessed with a blissful childhood in a semi-rural idyll, where tolerance was the order of the day.

The tone of Czechoslovakian politics throughout Herwig's childhood was set by Masaryk. The country's first leader had liberal leanings. He was a man of the world, having lived in the USA and married an American woman. His policies were inclusive and popular, and when Czechoslovakia held its first elections in 1920, he won comfortably to become the country's first elected president.

Partly as a legacy of the Austro-Hungarian Empire, Czechoslovakia had Polish and Hungarian minorities as well as its German-speaking population. The dominant religion was Catholic, but there was religious tolerance too. "Ninety five percent of the population was Catholic," said Herwig, "so of course, in school, the religion we were taught was Catholic. There were some Protestants and Jews, but there were no problems linked to religion at all. Since information about what was going on in other countries was practically non-existent we didn't know what was happening in Germany until 1938, when we became part of Germany."

Herwig's parents were not particularly devout, but the church nevertheless played an important role in his childhood. "My father never went to church, he was very liberal and the same was true of my mother," said Herwig. "My mother used to say that when she went to church, the saints all shook their heads because she went so rarely, but of course it was automatic as a young boy that I had to follow the religious ceremonies that were going on in town. From the gymnasium, I had to go to church every Sunday, but that felt more like a musical celebration than a religious one." Herwig's job in church was to operate the bellows that provided pressure for the church organ. His abiding memory from those Sundays is of the theatrical beauty of it all. "I was very impressed by the Catholic mass," he recalled. "I learned later that Richard Wagner tried to unify speech, music and spectacle in one cultural piece, the Opera, but this had already been achieved in the Catholic mass. I wonder if Wagner learned from that? In Church, I discovered what a real unified cultural event is. I was influenced religiously by the ceremonies of the Catholic Church, but not by my parents, and I must say that it took a certain effort to liberate myself from the influence of the Church, to develop a more liberal attitude."

Herwig does not remember living in a single close-knit family unit as a young child, but rather two. Apart from having a ten-year age gap, his parents were also very different people. By the time of Herwig's first memories, they had amicably divorced, and both had remarried. Neither had any more children, so Herwig was the only child of two families, with all its attendant consequences. "I had four parents and both families loved me very much," he recalled, "in fact I was completely spoiled."

For many years, the two families remained in Landskron, Franz choosing to remain in an apartment in the former country estate of Annahof on the edge of the town, where Herwig was born, while Margarethe and her second husband, Adalbert Hartmann, favoured the bustle of the town centre. "I think that was one reason for the divorce," said Herwig. "My mother did not like the rural environment, but rather the modern environment of the town."

Herwig divided his time between the two, and enjoyed the benefits of both. "In the divorce it was decided that I should stay with my mother during the week, and with my father at the weekend. At first, vacations were also with my father," he explained. "After my fourteenth birthday, this was flipped and I was supposed to stay with my father and spend weekends and vacations with my mother." It was an arrangement that suited him quite well.

He got to taste the pleasures of both the countryside, where the snowy winters allowed him to learn how to ski, and the town, where he could enjoy amateur theatre and orchestra, and even experience the new-fangled innovation of cinema. Although impressive with its four imposing wings, Annahof belonged to an earlier century. It had no running water, for example, and a restaurant that was located in one wing chilled its beer in the warm summer months with ice hewn from a local lake in winter. "For us youngsters it was always a big event when they brought the ice by horse cart to the estate," said Herwig. His mother's new husband, a lawyer, preferred an apartment in the town centre with all mod cons including running water and even a WC.

Gruss aus Landskron vom Annahof

Fig. 1.2 Herwig's childhood home, Annahof, at around 1900. Herwig was born in the ground floor room on left (Karel Uhlíř collection, reproduced courtesy of the Městské Muzeum, Lanškroun ©Městské Muzeum, All rights reserved)

Franz's second wife, Friederike Culik, was a piano teacher. "She loved music very much, and that influenced me greatly," said Herwig. "She taught me how to play the piano and through her, I too learned to love music." But the kind of music that the young Herwig grew to enjoy at Annahof was not the kind that would serve him in Landskron. "There was a lay orchestra in the town, and since they didn't need a pianist, I learned to play the double bass because bassists were very rare," he said. "So there I was, a little boy running around carrying this big bass instrument with me. The combination of the piano and the bass taught me to appreciate music not only from the point of view of chamber music but from the point of view of orchestra as well, so that was an important aspect of my youth."

School in Landskron was still organised along the lines of the educational system of the Austro-Hungarian Empire. Franz Schopper had qualified as a teacher of German and Latin before the Great War. After serving on the eastern front, spending much of the war as a Russian prisoner, he was assigned a post as a professor at the Landskron gymnasium. "At that time the practice was to go to elementary school for five years," said Herwig. "After that a decision was taken on whether you would go to the gymnasium or continue what was called popular school, *Volksschule*." Only about ten percent of children went to the gymnasium, which gave them the possibility of going on to university, and Herwig was one of them. Not only that, he did so well at elementary school that he was promoted to the gymnasium a year early, at the end of his fourth year. "This had the effect that I joined a class where all my classmates

Fig. 1.3 Herwig skiing in the Adlergebirge (Eagle mountains, or Orlické hory in Czech) near Grulich (Kràliky in Czech) in the 1940s. A poor mountain village at the time, Herwig's entertainment through the long winter evenings consisted of learning the craft of hand weaving (Herwig Schopper's personal collection. ©Herwig Schopper, All rights reserved)

were one year older than me," he recalled. "That became very important later when the war started."

Herwig's gymnasium was what was known as a *real gymnasium*, which had the distinction that it focused on science, an important factor in Herwig's development. Although Latin remained on the curriculum, ancient Greek made way for maths, physics, chemistry and biology. Two modern languages were also a requirement, Czech and English. "I must say I was very fortunate to have very good teachers," said Herwig, "especially in mathematics and physics. By the Maturity I'd learned algebra, differentiation and integration, and that helped me very much later." The Maturity was the name of the school-leaving exam.

Fig. 1.4 The gymnasium at Landskron (now Lanškroun) where Herwig's father Franz taught, and Herwig went to school (Palickap, CC BY-SA 4.0 DEED)

Another teacher that Herwig remembers fondly is his English teacher, a Jewish woman who had spent time in England before teaching in Landskron. "At that time, the tradition when learning English was not to speak it because there was no opportunity to practice: nobody spoke English, there were no foreigners around, so one learned English in order to read Shakespeare," said Herwig. "Fortunately, this teacher, having been in England, also taught us to speak English. That saved me later. In fact it was essential for me after the war."

The academic home life that Herwig had with his father helped to stimulate his interest in science, or more specifically in engineering. Other professors from the school were regular guests at Annahof, and thrilling home experiments were a regular occurrence. "When my chemistry and physics teacher visited us," recalled Herwig, "he helped to stimulate my interest in physics in a way that would be unimaginable today. For instance, he once brought to our home a bottle of mercury and I remember he put the mercury on the kitchen table and I could play with it with my fingers, bringing together bigger and bigger bubbles. It would be impossible today to have children play around with mercury like that, but it doesn't seem to have done me any harm." Over time, Herwig formulated a plan to combine his interest in technical subjects with his desire to travel: he would become an engineer on a ship.

His later change of heart had nothing to do with the school, and everything to do with his maternal grandfather, Franz Stark, who had been a university professor in Trieste during the time of the Austro-Hungarian Empire. When he retired, Professor Stark and his wife set up a small hotel on the Adriatic coast at Laurana, today Lovran, near to Abbazia, today Opatija, a town that had been a favourite holiday destination of Empress Elisabeth, known as Sissi, and remained a fashionable destination for those fortunate enough to have the means to travel. Without this family connection, Herwig would not have been able to dream of Adriatic vacations, let alone mix with exotic hotel guests such as professors of physics from the far corners of the former empire. It was through such encounters that Herwig cultivated an interest in fundamental science: the notion that with just a few basic principles, all the wonders of nature can, in principle, be understood (see this chapter, In his own words).

In many ways, Herwig led a charmed life as a child. He was much loved by his two families, had a good education, opportunities to develop as a musician and the great fortune to be able to travel. He was raised under a liberally minded administration in a new country just beginning to find its place in the world. But against this idyllic backdrop, there were seismic changes underway in the wider world, and just across the border a newly assertive Germany was flexing its muscles.

Seismic Political Changes

In 1924, the year Herwig was born, Adolf Hitler was jailed following the Munich putsch of the previous year, although he would serve only a fraction of his sentence. Lenin's death had led to Joseph Stalin becoming leader of the Soviet Union. Italy had returned a fascist government to power, and in the USA, J. Edgar Hoover became head of the FBI. It was also the year that another dominant empire passed into history, as Mustafa Kemal Atatürk abolished the caliphate and the modern secular state of Turkey was born. By the time Herwig was in the gymnasium, Hitler had become a dominant force in German politics, exploiting the fear of otherness and capitalising on the fact that the German army had never surrendered at the end of the Great War. Weak politicians, he claimed, were the reason for the Armistice and the resulting punitive Treaty of Versailles. Hitler's heady brew of populism found fertile ground in Weimar Germany, and was matched by equivocation on the part of the victors of 1914–1918.

By 1935, the winds of change were picking up speed. When ill health forced Masaryk to retire from public life that year, his successor took a more strongly nationalistic approach to Czechoslovakia's minorities. By 1938, a proactive policy had been adopted to incorporate the various minorities into the young country. Landskron began its transition to Lanškroun. Some Czech families moved into town, and a Czech-language school was opened. Across the border in Germany, things were moving in a very different direction, and on 30 September 1938 at a conference in Munich, France, Great Britain and Italy yielded to German demands to cede the Sudetenland, a border region with a large German-speaking population to Germany.

Fig. 1.5 Lanškroun town hall, seen here in the 1980s, has changed little since Herwig lived in the town as a boy (Herwig Schopper's personal collection. ©Herwig Schopper, All rights reserved)

This included Landskron, which became German overnight. "Daladier and Chamberlain made me a German citizen," Herwig recalled, "but that didn't have much influence on my education, the gymnasium went on like before."

1938 was just the beginning of major changes in Herwig's life. It was the year that his mother moved to the town of Zwittau, today Svitavy, some 20 km away. That does not sound like much, but with the transport infrastructure of the day, it presented a major obstacle. "The rural roads were mostly just gravel, so to go to a neighbouring town even twenty or thirty kilometres away was an adventure," said Herwig. "It could take a whole day because the transport was horse-drawn carts, or you had to walk. There were a few taxis but they were very expensive and they were used only in very special circumstances, for instance if you got sick or had an accident and had to go to the nearest hospital. Only a few of the larger towns had asphalt roads and a bus service."

The following year, 1939, the unthinkable happened. The war to end all wars clearly proved not to have done so as the lights once more went out across the continent, and Europe again found itself in conflict. The consequence for Herwig was almost surreal. Having skipped a year of elementary school, he was a year younger than his classmates, all of whom were old enough to be conscripted by the Autumn of 1941, leaving the 17-year-old Herwig in a class of one. "My classmates were given an emergency Maturity qualification, they didn't have to go through the Maturity examinations," said Herwig. "They disappeared but I stayed and strangely enough, the whole teaching of that class continued with me as the only pupil." Things would get stranger still when it came to Herwig's Maturity examinations in February 1942. "Since my father was a teacher at my school, they were careful to make sure that I did not get any preferential treatment, so someone high up in the school hierarchy came from another town to be present in my Maturity examination." Herwig passed with distinction, guaranteeing himself access to university when such a thing became possible again. "I passed my examination in February'42, and one week later I was called to the *Reichsarbeitsdienst*—the Reich's compulsory labour service."

In His Own Words

The Wonders of Physics

"An important point that was essential in my life concerns one of my mother's parents, who was a university professor in the old Austria before the First World War. He was a professor at the University of Trieste in Italy. When he retired after the war he bought a villa in a place which is today called Lovran, at the time it was Laurana, which is close to Abbazia, which was, in the old Austria, one of the favourite vacation places for Austrian aristocracy and bourgeoisie.

Abbazia is now called Opatija, and it is still a popular resort. In the old Austria, it was famous because the Austrian Queen Sissi—Elisabeth—vacationed there several times. My grandmother, who came from Landskron, was a devoted mother and she also liked to cook for people, so she convinced my grandfather to open a hotel. They bought a villa at Laurana and converted it into a flourishing hotel where I spent my summer vacations.

Many people from Vienna, Budapest and other cities from the old Empire came to that hotel, and I met two in particular. One was a professor at the University of Budapest, Tibor Neugebauer, and the other was a professor called Plotnikov from the University of Belgrade. They were both physicists. Before the war it was very hard for people to go on vacation to that kind of place, only very rich or very well-situated people could do it. Very few people in Landskron could afford to take a vacation in Italy, but thanks to that grandfather in Trieste and his hotel in Abbazia, I spent most of my summer vacations, whether I was formally with my father or with my mother, on the Adriatic coast.

I remember that I was three years old when I spent my first vacation there. From the beginning I was always interested in science—well science was not so well known—so I was interested in engineering really. At first, I wanted to become an engineer but I also liked to travel, so I wanted to become an engineer on a ship. That was my dream when I was a youngster, but I had no idea at that time what physics was.

During these vacations in Laurana, where I met these two gentlemen sitting by the sea, I learned about physics. After swimming in the Adriatic, I would listen to them—their conversation was all in German because they came from the old Austria. One day they would be talking about how butterflies can fly, the next evening they would be looking at the sky and discussing why the stars are burning, where they get their energy from and things like that. I was so impressed that these gentlemen, these physicists, could discuss such completely different phenomena with such authority. One day they discussed how the wind produces waves in the sea and so on. I was so impressed that they could discuss completely different phenomena based on just a few basic principles. I didn't yet understand those principles, but I did understand that this was the basis of physics.

That was the first time I came into contact with physics, and it made me change my mind about becoming an engineer. I decided to become a physicist, so that was an important milestone in my life. Thanks to my grandfather in Trieste, I could spend my vacations in Italy, and there I met the first physicists of my life, before that, I didn't know what a physicist was.

In Laurana I had another experience that is an important element for most young people—my first great love. I met an Italian girl, Nedda Ferri, she was about my age and we fell deeply in love even though she did not speak any German and my Italian was rather rudimentary. Every year we would long for the summer vacation to come around, and we spent every possible moment together. Our parents on both sides seemed very happy about this situation. According to the moral rules of the time, we respected all proper limits in a way that seems hardly believable in present times. Since we were separated for most of the year, we exchanged many letters and I learned not only to speak Italian, but also to write it. For her part, Nedda learned German and many years later became a German language school teacher at Meran in the South Tyrol. Our meetings, albeit very short, continued until I became a soldier. But they could have changed my whole life as I will recount later. Eventually we lost contact, as did so many during the war. We met again about 50 years later, and remained friends until she passed away in 2012."

Private Life

"Although my parents divorced, they did so amicably, and both gave me all the love and care a young boy needs. It also seems that they bequeathed me good genes. I have lived for almost a century without ever being seriously ill. In all my long career, I have never missed a single day of work due to illness, and the only days I had to spend in hospital were due to skiing accidents. My lifestyle probably played a part too. I've never eaten or drunk to excess. I've never smoked, not because of a strong character, but just because I disliked the stinging in my nose and eyes. And I always liked sport. At school I played football, and then took up tennis and skiing. When we could afford it, we built a swimming pool that I still use to this day to keep myself as fit as possible. My children learned to swim at almost the same time as they learned to walk.

Another important element in my life has been my love for music, in particular playing the piano. I learned when I was young, mainly taught by my father's second wife, but I had to interrupt my playing for many years during and after the war. On occasion, I enjoyed performing chamber music, accompanying a violinist or quartets. I never approached a professional level, but I liked very much to use recordings that featured an orchestra playing a concerto with the piano removed, so I could play the part myself. In this way, I played many of the Mozart concertos and sometimes even dared to attempt some Beethoven, or even Chopin, although I have to confess that I used a computer program to slow down the speed without changing the pitch. My only public performance was at a little meeting at CERN when I was asked to play for staff members who had 25 years of service.

My private life was multifaceted, having had effectively two pairs of parents, being a soldier in the war, a displaced person and a prisoner of war, and having to find my own way afterwards alone. On a few occasions fate took the lead in determining the course of my life, as I've mentioned in earlier chapters. I learned early on that it is an illusion to believe that our lives are exclusively determined by our planning and our actions. Luck plays an important part.

I owe much of the happiness of my personal life, and also my professional career, to a very long and happy marriage to Ingeborg Schopper (b. Stieler), which lasted 60 years; an increasingly rare period for couples to be married these days. Of course, our marriage had some ups and downs but my wife and later my children, Doris and Andreas, were always accommodating when my career took us to new places around the world—different countries, languages and schools. I have been very lucky in this respect, and I value how important it is to have a family with whom to exchange affection, dedication and love. If I have one regret, it is that I could not devote more time to them.

To my great sadness, my wife passed away more than a decade ago and although my children take great care of me, I have also the good luck to find a partner, Ingrid Krähe, with whom I share many values and beliefs and who brightens my life today.

My professional career was marked by a lot of variety. I never stayed longer than about eight years in any one place. Instead, I moved through a changing social and professional environment, which gave me the opportunity to get to know different actors in society rather intimately, from soldiers and students to world leaders. I regularly joined new scientific environments, in which I had to build a reputation from scratch, whatever I might have done before. This meant gaining recognition by a new group of peers, which was sometimes rather difficult.

Fig. 1.6 As one of Lanškroun's more famous sons, Herwig was honoured with his likeness being featured in the town's nativity scene, developed between 2000 and 2011. The scene is now on display in the Městské Muzeum, Lanškroun (reproduced courtesy of the Městské Muzeum, Lanškroun. ©Městské Muzeum, All rights reserved)

Curiosity has been central to my life, at least consciously, ever since I eavesdropped on the conversations of physicists on the beach in Abazzia as a child at my grandparents' hotel. As I developed into a working scientist, I began my career trying to understand what is behind the laws of nature that govern all natural sciences. These efforts induced me to look beyond the sciences to the motivations that drive the actions of people, and to comprehend the ideas that shape economics, politics and history."

Chapter 2
The War Years

With the annexation of the Sudetenland in 1938, swiftly followed by German occupation of all the Czech lands in 1939, Herwig's childhood idyll came to a brutal end. The unstable peace that had prevailed since the largely German-speaking borderlands had been assigned to the newly established Czechoslovakia at the end of the First World War had started to crumble, with the rise of German nationalism translating into electoral success in 1935. The young Herwig had been sheltered from these developments, but by the time he began his final year at the Landskron gymnasium, an air of anxiety was spreading across Europe. Herwig had little doubt what lay in store for him following his graduation. It was therefore with some trepidation that he packed his bags in March 1942 and left for a small village near Breslau.

Herwig was on his way to join the RAD, or *Reichsarbeitsdienst* to give it its full name, a labour service established by the Nazi Party in 1935, nominally to combat youth unemployment, but in reality as a means of indoctrination and a precursor to military training. "In the camp, we got a kind of training that was more military than anything else," he recalled. "Instead of training with a rifle, we had to use a spade." The work was physically exacting, but the real challenge was psychological: "They immediately started to try to break our will so we would learn to be obedient to orders, that was the hardest thing."

In the early days of the RAD, Germany maintained a pretence that it was a voluntary organisation, but by 1942, the days of pretence were over. The RAD became obligatory for every young German man. Besides providing a cheap labour force, the RAD became established as a pre-military training organisation. After they'd been at the camp for a few weeks, the young conscripts had a visit from the Waffen SS, the military wing of the notorious Nazi organisation. "We were told that it's the greatest honour for a young German guy to be called to this Waffen SS and defend the country," said Herwig. "If you said 'no,' you were sent to the front in a battle where your life expectancy was just a few weeks. Luckily for me, my health was not good enough to be recruited thanks to my spectacles: so I thank my short-sightedness for my narrow escape!."

© The Author(s) 2024
H. Schopper and J. Gillies, *Herwig Schopper*, Springer Biographies,
https://doi.org/10.1007/978-3-031-51042-7_2

After the SS visit, Herwig was transferred to Thuringia and put to work in a disused salt mine that had been converted into a munitions factory and storage depot. Some of these mines were used to store Nazi gold and stolen works of art [1], but not this one: it was a munitions repository for Rommel's *Afrika Korps*. Herwig's job there involved descending to around 700 m below ground and filling steel grenade cases with explosives for eight hours per day. On top of that, it took an hour to march to the mine and another hour to return, and in the evening there was work to be done to maintain the camp. "That was very hard, boring and tiring work, but I survived."

Basic Military Training

By August, his time with the RAD was over and Herwig had a few days at home in Landskron before being conscripted to the military. "Fortunately," he recalled, "they asked me what part of the army I wanted to join, and since I wanted to stay in touch as much as I could with science and physics, I thought the air force would be the place with the highest intellectual level." He was assigned to the *Nachrichtendienst*, the signals corps of the air force, and in September 1942, he was put on a train to the small town of Auxonne on the banks of the river Saône in Burgundy. It was to be the calm before the storm.

Fig. 2.1 Conscripted. Herwig and his schoolfriend Kurt Hayek leave Landskron. Herwig was on his way to Auxonne in occupied France for military training (Herwig Schopper's personal collection. ©Herwig Schopper, All rights reserved)

On arrival in Burgundy, Herwig discovered the unmistakable contours of a Vauban-fortified town, and moved into its Napoleonic barracks, which seemed to have changed little over the intervening years. "We were installed in sleeping halls, about fifty people to a hall, and the sanitary conditions were still the same as in 1788 when the place was built, so it was not paradise, and of course we went through the normal drill, this time not with a spade but with a rifle." Life for a member of the signals corps involved a lot of physical exercise and training, but there was also an academic component. Lectures brought the new recruits up to speed with the latest information technologies, telephony and Morse code, and as such technology was built on science, there were courses in physics too. "That was very amusing because the teacher was a young Sergeant who had no idea about physics," said Herwig.

Herwig's fellow recruits were not only young men with their Maturity diplomas freshly awarded, but also some engineers who had worked for Siemens before the war. It all added up to a student body more knowledgeable than the teacher, with sometimes amusing, if sobering, consequences. "One day we discussed the direction of electric current, and he told us it goes from plus to minus, so we said,' but you told us that the electrons are charged negatively, so they should go from minus to plus,' he was completely confused," Herwig remembered. The students may have had fun teasing their teacher, but at the end of the day, it was always clear where the power lay, and what the future held in store for the young conscripts. "It doesn't matter what direction the electric current flows," said the Sergeant. "The only thing that matters is that if a General comes to inspect our company, the answers you give must be uniform, so I order that the current goes from plus to minus!."

After several weeks confined to the barracks of Auxonne, the conscripts were eventually allowed out at weekends. For Herwig, this had the double advantage of providing some real rifle practice—food was in short supply, and they hunted wild boar in the local woods—and allowing him to become re-acquainted with his love of music. Among Auxonne's architectural treasures is a gothic church, which was equipped with a fine organ. On Sunday afternoons, Herwig and a fellow conscript, a talented musician, would go to the church, and Herwig would be treated to concerts just for him. "That was the best part of my experience as a conscript," he remembered.

As his basic training neared its conclusion, Herwig had to decide whether to be a regular soldier, or to volunteer for training as a reserve officer. "Since I was afraid that as a simple soldier I would be somewhere at the front for years, completely out of touch with intellectual life, I opted for officer training" he explained. That decision set the rhythm for Herwig's entire war: he had six months of training interspersed with periods of duty at the front to look forward to. Much later, Herwig had the opportunity to revisit Auxonne, and he found things much as he remembered them: "When I visited Auxonne as a tourist in 2021, I saw that the barracks had been modernised somewhat. The town was still the same typical provincial French town, but a harbour for boat trips had been created on the Saône where I used to go for long walks along the banks of the beautiful river."

Fig. 2.2 In 2021, Herwig returned to Auxonne to visit the barracks he was sent to for military training in 1942. The area is known as the Quartier Bonaparte, since it is there in 1788 that a young future Emperor of France learned his military craft. Today, the barracks are home to the 511th Transport regiment of the French army (Herwig Schopper's personal collection. ©Herwig Schopper, All rights reserved)

After six months in Auxonne, just before he was due to start active duty, a chain of events began to unfold that could have changed Herwig's war, and indeed his life, beyond recognition. As well as setting him on course for a career in physics, Herwig's pre-war holidays with his grandparents on the Adriatic coast had another dimension. "I had fallen in love with Nedda Ferri," he recalled, "we were very close, and even considered marriage."

With the emergence of war all such plans were put on hold, but when Nedda's father learned that Herwig had been conscripted, he tried to get him assigned to Italy to serve as a link person between the Italian and German armies. It almost worked, but the transfer request came too late. "It arrived two days after I'd left for the Russian front," explains Herwig. "If it had arrived on time, my life would probably have been completely different."

The Eastern Front

It was in March 1943 that Herwig found himself at the front for the first time. He had received the order to travel to Russia to join a unit responsible for telephone services, and was stationed between Minsk and Vitebsk in a tiny village called Ust Dolissi. "Our job was to ensure that the telephone lines connecting two division headquarters behind the front remained operational all the times and we had to make tests continuously." The region was heavily forested, and provided ideal terrain for resistance partisans to ply their trade. So although the signals corps was not a fighting company, Herwig regularly found himself in the firing line. "The partisans came during the night and cut the lines and even the telephone masts," he recalled. "They knew that within a few minutes we would be there to repair them, and they were waiting with their machine guns. The moment we climbed up the masts to fix the lines, they shot at us. I must admit, this was one of the most dangerous and most unpleasant experiences I had during the war." Herwig learned how to turn trees into telegraph poles and to scale them with iron crampons attached to his feet to string the telephone wires, all the while with a rifle on his back in case of attacks by the partisans. To make things more unreal, Herwig's unit was billeted in a local village and the young German soldiers had to interact with the villagers. "It was probably the same people we were dealing with during daytime to buy food that were shooting at us at night."

Despite the shock of finding himself in a war zone for the first time, Herwig nevertheless found ways to advance his knowledge of physics. Before leaving home, he'd signed up for courses at the Technical University of Prague, and despite the war, the university honoured the commitment it had made to its new students. "They had a group there that was taking care of soldiers by sending them exercises in physics, which we had to solve, and send back to Prague. They corrected them and sent the work back. So I was sitting there in the dark blockhouses infested by all kinds of insects, solving physics problems, while watching the telephone line of course, day and night. We constantly had to test whether the line was working, so in between calling the next post every quarter of an hour, I was doing physics exercises. I was very grateful to the tutors at Prague who I never met."

The telephone line itself also proved to be a source of physics learning. "There I learned a lot about telephone technology. For instance, if you look at telephone lines even today, you will notice that if there are two telephone wires, they don't just go in parallel, but they rotate about each other. This is a way to reduce the induction between the telephone line and the earth, and to improve the quality of the communication."

After six months, Herwig's first stint of active duty came to an end, and he was ordered to attend the Luftwaffe's Military Academy in Halle. It was now the late summer of 1943, the war had passed its mid-point, and the ascendancy of Germany and the Axis powers was waning. In the Pacific arena, the Allied victory at the battle of Midway in 1942 had marked a turning point, and by the time Herwig arrived in Halle, Allied victories were mounting up. American troops had arrived in Britain. The stalemate in North Africa had resolved in the allies' favour. In Russia, the German army discovered the harsh realities of the Soviet winter as they were defeated at Stalingrad, and with the defeat of Italy, Germany lost an ally. By the end of the year, the allies would be planning their return to continental Europe.

Against this backdrop, Herwig's time in Halle took a surreal turn. "In the middle of the war you would have thought that everything would be concentrated on military training," said Herwig. "What is hard to believe looking back is that they thought that future reserve officers should not only learn military matters but should also get general social education, and they thought a future officer should learn how to dance! It's hard to believe, but in the middle of the war we had dancing courses in the evening with girls of Halle. Completely crazy!"

It wasn't all foxtrots and quicksteps, though, there was also training in the more serious business of the signals corps. "In Halle we got a really good training in signals technologies. I had to learn Morse code to send coded messages. The Morse equipment was a key and headphones, and I had to learn to send and pick up two hundred letters a minute, which was not easy. It was hard training and I've never forgotten it. Even today I can communicate in Morse code, although only at much reduced speed."

Later in life, Herwig had the opportunity to reflect on his time in Halle. In 1967, he was appointed a member of the Leopoldina [2], Germany's oldest continuously existing learned academy, which was established in 1652 as the *Academia Naturae Curiosorum*, and took the name of Emperor Leopold I in 1687. After having several homes, the Leopoldina finally settled in Halle in 1878. In 2008, it became Germany's National Academy of Sciences, but at the time Herwig became a member, it was still behind the iron curtain.

After the war, Halle found itself in the German Democratic Republic, East Germany, and under normal circumstances, it was impossible for West Germans to go there. As a member of the Leopoldina, however, Herwig could go and this gave him the rare opportunity to foster contacts between the physicists of East and West Germany. As an active member of the academy, he frequently had business there, and on one of his visits after German unification, he decided to find out what had happened to the old Luftwaffe barracks, just across the river Saale from the town. "I was very surprised to find the old buildings still there," he recalled. "They had not been destroyed, they had been transformed into a university campus, and in the building where I had spent time as a soldier, there was the University's physics institute. A much better use of the space."

After his six months in Halle, he was sent back to the Russian front, although this time having passed through the academy and emerged with the rank of *Fahnenjunker Unteroffizier*, which translates as Officer Cadet Corporal. By now, it was 1944, the year in which Allied troops would return to France through the beaches of Normandy.

This time, Herwig was again posted to the northern part of the Eastern Front, where he joined a company that had the job of providing regular weather reports every half hour, and guiding Luftwaffe planes supporting the battle on the ground from the air. The company was composed of several small groups of about a dozen people each, stationed a few kilometres behind the lines and controlled from company headquarters at Riga-Strand, today known as Jūrmala, a seaside town on the Bay of Riga.

It was first to the HQ that Herwig was sent, and it was there that his increasingly surreal war became even more strange. He found himself as a member of the officers' club in a beautiful Baltic coast resort, and he was expected to wine and dine with his fellow officers every night, while a Soviet offensive was raging just a few hundred kilometres to the east. "I was not only entitled, but I was obliged to participate in the officers' daily life," he recalled. "Among other duties, I had to eat in the officers' special restaurant, a kind of club. The officer in charge of the company, a Major, was a righteous man, as I learned later in a serious situation, but my first impression was that he liked to drink."

Every evening, Herwig was invited to join the officers at the club for drinks. They were an ageing group, and glad of some younger company. "I have nothing against drinking within limits," said Herwig, "but I have never got drunk in my life. When I drink too much I get tired and don't feel well, I do not get cheerful and happy." So after a few evenings of drinking, Herwig asked the Major whether he could leave and go to bed, a serious breach of protocol, as it turned out, but the Major replied, "Yes, you can go to bed."

The following day, Herwig got the order to join one of the groups at the front, just two kilometres behind the lines, and his war was jolted back into stark reality. "I was again billeted in a little village, but this time I found myself responsible for a group of twelve people, some of them forty or fifty years old, married with children," he explained. As a young man of barely twenty years of age, Herwig found himself in charge of a group of men mostly his senior. "This was a typical experience for my generation, having to take responsibility at a very young age, and with very little experience, for the lives of other people, even concerning life and death. Such a commitment would be impossible nowadays, and the responsibility was not trivial."

As a member of the signals corps, Herwig's was not a combat role. His unit provided weather reports to Riga-Strand using Morse code every half hour. The messages sent by wireless consisting of short and long strikes had to be rigorously standardised, because the enemy was listening, and was on the lookout for anything out of the ordinary that might suggest movement on the German side. Herwig's unit did not have the famous Enigma machines at their disposal: they had to code by hand using a manual system known as the *Doppelkastenschlüssel*, or Two-Box Cipher, which was easier to crack. But regardless of whether the enemy could crack the code or not, if any individual had a particular style of transmitting, enemy listeners could glean valuable information from the transmissions. For example, if an individual coder gave a signal that was just too long or too short, it would serve as a kind of signature, and the enemy could use that to identify whether he had moved along the front.

The penalty imposed by the German military for any departure from the rigorous code was prison, and there came a time when Herwig had to impose a sentence on one of his men. Despite the threat of sanctions, the temptation to use slight distortions of the code, a sort of code within a code, to convey messages to friends and comrades down the line was sometimes just too great. "Of course we liked very much to give secret personal information about how we were to comrades in other groups, and we'd agreed to just delaying a point or strike in the Morse code in a specific letter a little bit," explained Herwig. One day, when one of the people under Herwig's command did just that, Riga-Strand noticed, and sentenced the perpetrator to three days of prison. For Herwig, this presented a dilemma, as there was no prison two kilometres behind the lines. His solution was to confine the man to one of the vehicles that contained the transmission equipment, and he sentenced him to three straight days of weather reporting.

Herwig's own brush with military justice came later, and carried potentially more serious consequences. Although not a combat unit, all the men had been trained to fight, and if the situation at the front became tense, it was up to the superior officer to decide whether to fight or retreat, thereby protecting the valuable coding and transmitting equipment. The general brief was ambiguous: "you are responsible for several millions worth of equipment, which is partly secret and must not fall into the hands of the Russians, but do not retreat without serious reason." The day came in June 1944 when Herwig had to make a judgement call. "The Russians attacked and they got closer and closer," he remembered. "I established contact with the infantry, which were close to us, and I asked their advice about whether we should join them or not. The answer I got was: 'you do what you want, it's your decision.'" Herwig decided to pull back ten kilometres to protect the equipment. At the end of the day, it proved to be the right decision: the infantry saw off the attack, and the valuable equipment was safe. "Of course I had tried to get a decision from HQ but the wireless communication was not good and I had to take a decision without having received an order," he explained. "But that's not how the military authorities saw things: I was accused of retreating without order."

While this was happening on the Russian front, Allied forces were landing on the beaches of Normandy and making advances into France. Herwig's Morse code equipment kept him and his men abreast of developments, as the Luftwaffe's High Commander, Herman Göring exhorted his men to dig deep and fight to the victorious end, despite the setbacks being suffered by Germany on all fronts.

The need for redoubled effort did not stop the German military from putting Herwig on trial. The procedure was a complicated one, with all negotiations done by wireless. Herwig would spend hours decoding the messages coming in from company HQ—and sometimes the propaganda from Göring would even provide relief—but eventually a decision was taken. He was acquitted. The heavy-drinking Major in Riga-Strand stood up for Herwig, telling the court that there had been no time for an order to be issued, and that the young *Fahnenjunker Unteroffizier* had made the right decision under pressure.

This episode was testing for Herwig in another way too. "At that time, I really saw what war is," he said. "We were so close to the infantry that was defending our position that when we came out in the morning we saw all the dead bodies that had been killed during the night lying in the streets, it was horrifying."

The End of the War

Freshly acquitted, Herwig was sent back to Germany in September 1944 for another stint of training, this time in Berlin at an Officers' Academy in the suburb of Kladow. This was very different from the training in Halle. Although it involved tough, physical, military training, this time it also had a strong intellectual component. Training was marked by two, very different, trainers: one a tank-busting and highly decorated officer in the Luftwaffe signals corps, the other a highly educated intellectual. He was a member of the General Staff, a *Generalstabsoffizier*, and as such a member of the intellectual elite of the German army, recognisable by a red stripe on his trousers harking back to the uniforms of the previous century. The General Staff had been responsible for the Prussian, and then the German, military's strategic planning since its foundation in the early nineteenth century, although under Hitler their power had waned, because Hitler considered himself a superior strategist. Members of the General Staff were frequently intellectuals, and often dissenters. Herwig suspected that this might be the case for his new teacher.

Opposition to Hitler, in the military and among civilians, pre-dated the war, and came to a head in 1944 with the failed 20 July plot by Claus von Stauffenberg and his supporters to assassinate the Führer at the Wolf's Lair complex in East Prussia. Whether Herwig's teacher had been involved with that plot, he would never know, but it was clear to anyone looking for the signs that this *Generalstabsoffizier* was no Nazi. "He was a very educated man, and he not only had to teach us tactics but also history. We realised immediately that he was an anti-Nazi, but of course he couldn't give any public indication of that because he would have lost his life immediately. In a very sophisticated intelligent way he explained to us all the mistakes of Nazi propaganda and errors of Nazi ideology, and he did that by comparing implicitly the Nazi mentality to communist mentality pretending the two were opposite. But of course, we were clever enough to compare the two and understand that all the negative things he told us about communist ideology applied also to the Nazi mentality. So in that very intelligent, clever and sophisticated way he opened our eyes or strengthened already existing tendencies." Not that Herwig would ever have discussed his political feelings with his fellow trainee officers. With the atmosphere of oppression and distrust that reigned, you just never knew who believed what. "It was impossible to discuss politics because, even if you tell your best friends that you doubt the final victory—Endsieg it was called—you were considered a defeatist. You got a war trial and you were accused of the crime of destroying the morale of the army and could be sentenced to death."

So Herwig kept his opinions about what was really going on in the *Generalstabsoffizier's* head to himself, and dutifully followed his training. The other notable instructor could not have been more different. "Our Lieutenant had fought Russian tanks with so-called hollow charges, which had been developed to break the armour of tanks. To use them, a soldier had to run up to the tank and stick the hollow charge equipped with magnets on to it. It would then fire through the blinding and destroy the tank. When somebody managed to do that he got a special medal for anti-tank fighting, and this Lieutenant had several of these medals. When we complained about the hard training he said: 'look, it's not just to make life hard for you, it's in your own interest, because at war you have to follow orders otherwise your chances to survive are very small. We have to educate you in such a way that you follow orders without discussion.' Such a strategy was psychologically not easy to accept. But it was war and there was no alternative." Despite the news coming in from all fronts, the general feeling among the soldiers was that if Germany surrendered to the allies, the German military would soon be fighting alongside them, against the Russians. So Herwig took his Lieutenant's words very seriously.

While in Berlin, Herwig was also assigned to one of the most advanced telecommunications groups in Germany, or indeed anywhere. In the nineteenth century, Heinrich Hertz had shown that metallic objects could reflect radio waves. Although he did not identify any practical use for this observation, it was later deployed in the 1904 invention by Christian Hülsmeyer of a device called the Telemobiloskop, which helped to prevent collisions between ships at sea. In the Second World War, it was developed further into what we now know as radar, short for radio detection and ranging: a precision tool for pinpointing metallic objects in the sky.

The German implementation of this took the form of the Würzburg-Riesen, large 7.5 m diameter radar mirrors that were deployed in the defence of towns when the Allied bombardment began. "The defence units had two such radar facilities available to them. In the beginning, when the bombardment of Germany by British bombers started, attacks were carried out by individual bombers, or small groups of two or three planes. So one of these radars could locate the enemy bomber while the other picked up a German night fighter. The German fighter was guided to be placed behind the British bomber," explained Herwig. "This was achieved with the help of a glass table with two projectors from below, projecting a blue point which was the German fighter and a red point which was the enemy. An officer talking to the German pilot could guide the German fighter behind the bombing plane, and at just the right moment the German pilot would get the order 'pauke-pauke' to start his machine gun and shoot at the bomber without even seeing it. This was the most advanced technology at that time and it's how I learned about radar."

The success of the Würzburg-Riesen was to be short lived: as the Luftwaffe lost its domination of the German air space, huge squadrons of bombers would appear in the skies over Berlin every night at 10.00 p.m. "It was so punctual you could set your watch accordingly," recalled Herwig. Individual fighters could do little to resist, and the ground-based defences fared little better. Every morning, Berlin would wake to more devastation from the overnight raid. "They were not dropping so many incendiary bombs, but bombs that created explosions by air pressure, with the consequence that the buildings were not damaged by fire, but by pressure. Most of the windows were broken, and sometimes the ceilings fell down, but the furniture was still intact." Here, Herwig's military training was radically transformed, as the officer students were detailed to salvage what they could. "In the morning, we were transported to the centre of Berlin and we helped to clean up the damaged apartments and take out the furniture that was still intact. At the time, I learned to hate pianos because I had to transport a lot of pianos from the third floor down to the ground. It was very sad, because when we entered Berlin every morning—one morning the Opera was destroyed, the next morning we saw a museum was in ruins. Next morning a church was in ruins—a very depressing experience. I lived there through the bombing expecting that one night our barracks would also be bombed, but the military camp was never attacked only the town of Berlin itself."

By the end of 1944, Herwig graduated from the Military Academy as *Fahnen-junker Leutnant*, a candidate Lieutenant who after a short period of service would automatically become a Lieutenant. He was dispatched to the Western Front, with no possibility to visit his family on the way. As 1944 turned to 1945, things were becoming increasingly desperate for Germany as the war entered its final phase in Europe.

Herwig joined a Divisional HQ tasked with guiding German fighter planes in the battles now raging across northern France, but by this time, the Luftwaffe was a spent force. "The German fighters were useless because there was no fuel anymore. They were fixed to the ground and there wasn't much we could do. As a last attempt an offensive in the Ardennes mountains was started by the German troops and we were sent there to help the infantry. But after a few days we were recalled to the Rhine, to defend the bridge of Remagen."

The Rhine was a very important psychological landmark for Germany. If the allies crossed it, they would be on German soil, and Germany threw everything it could at preventing a crossing. Herwig was assigned the particularly grizzly task of recruiting pilots for kamikaze raids to destroy the bridge before the advancing allies could use it to establish a bridgehead on the eastern side. "One of my tasks was to make the connection with the pilots, to find pilots who would volunteer to do that. Quite a number were prepared to, but there was not enough fuel anymore and the bridge was taken by the Allies. We were asked to join the infantry to defend it with our rifles, a ridiculous enterprise! But before we could even start, the Allied forces had crossed the bridge and formed a bridgehead." A hugely symbolic victory had been achieved. "Many Germans, including myself, were led to the belief that the Western Allies would join with Germany in the conflict with Russia."

With the Rhine breached, the emphasis was given by Hitler to defend Berlin from the approaching Russian troops. Herwig was ordered to the capital with a few men under his command. "I was charged with a little group including three trucks and a few people to find our way from the Rhine to Berlin and report to a headquarters there, but Germany was in chaos. How could we get from the Rhine to Berlin? Somehow, we managed using somewhat unusual tactics. For example, a member of my group managed to rustle up a sack of green coffee, which was a much-appreciated currency at that time. We loaded it onto one of the trucks and started on our way to Berlin. I was in command, and I had my own jeep with a driver and we headed this little convoy, a jeep and three trucks. Of course, the allied air forces controlled the airspace above Germany completely, day and night. To move during daylight was very dangerous. At night, it was not much better and it was also harder for us to find our way through unknown regions, so most of the time we tried to advance in daylight. To get food we usually used the coffee to buy on the black market."

One day, Herwig and his small convoy were driving through North Rhine-Westphalia in broad daylight, and they were spotted by a squadron of Spitfires. "We left the trucks on the road and spread out into nearby fields, the Spitfires came down firing at us. They came so close I could almost look the pilots in the eyes. But, we were lucky. None of us was killed or wounded. They destroyed one of our trucks but my jeep and the other trucks were still intact and we continued to Berlin."

They arrived in March 1945 to find total chaos, and discover that Hitler's strategy had changed again. Nazi Germany was down to its last two strongholds, and the high command was determined to fight to the last. "One stronghold was in the Alps, where the commander was Göring, and another stronghold was supposed to be created in the north of Germany, in Schleswig–Holstein near the Danish border. There Admiral Dönitz was put in command. So after a day in Berlin we suddenly got the order to join the stronghold in Schleswig–Holstein."

By this time, it was clear that the war was effectively over. His marching order for Schleswig–Holstein was a very welcome document because Herwig had already reached the conclusion that it would be better to be captured by the western allies than by Russia. Being sent west not only allowed him to avoid summary execution by the German army as a deserter if he'd gone it alone, it also allowed him to move across the Elbe, where everyone expected the line of control between Russia and the western allies to be drawn. Nobody wanted to be a Russian prisoner of war.

After a quick visit to Frederik the Great's testimony to another, more enlightened period of German history, the palace at Sanssouci, Herwig and his driver set off for the Elbe, and onwards to Schleswig–Holstein (see this chapter, In his own words: a visit to Sanssouci). By the time they arrived at the end of April 1945, the war was all over bar the shouting. Hitler was dead, although that was not widely known, and Germany was on the point of unconditional surrender, signed by Alfred Jodl, Chief of the Operations Staff of the German Armed Forces High Command, in Reims on

7 May, with a ceasefire to begin at midnight. Jodl had one final attempt to persuade the allies to join Germany in fighting the Russians, but Eisenhower had none of it. As if to underline the finality of the German defeat, a second act of surrender was signed two days later by Field Marshal Wilhelm Keitel, the supreme commander of all German forces. To this day, 8 May is marked as VE day in Western Europe, while Russia celebrates victory day on the 9th.

For Herwig, the Allied victory came as a relief. "Somehow, I have forgotten exactly how, I got a paper that said that I was dismissed from the German army, and I was not defecting. We threw the pistols that we were still carrying in the nearest lake, and moved on to Schleswig–Holstein, where we were caught by the British troops. I'd never seen a British person before in my life," Herwig recalled, "and it was a pleasure to discover that they treated us very reasonably. The problem was that they faced so many prisoners that they didn't know what to do with us. We were put in a camp, not a real camp, just an open field, and we were left to sit there on the grass for two weeks under a rainy sky. We were given a dozen biscuits every day, that was all we got for food, and some water to drink. Finally, the British were able to put us in a camp for prisoners, which was established in a former farm in Schleswig–Holstein. The only possession I had was my rucksack, nothing else, everything else was lost. After some time, the British military administration worked out a procedure to release prisoners to provide the labour desperately needed to start the reconstruction."

Prisoner release was a complex affair because the British were looking for SS soldiers, and perpetrators of the atrocities that had been laid bare as the allies advanced and liberated the concentration camps. Everyone had to be rigorously interviewed before release in an attempt to weed out the fugitives from the ordinary soldiers. The language barrier didn't help. "One day I learned that the British were looking for interpreters, so I told myself to be courageous and try," he recalled. "It was then that I had cause to be thankful to my English teacher at school, who taught us to converse, and not just to read Shakespeare. Among young Germans of my generation, that was a rare skill to possess, and it proved valuable to the British." Herwig plucked up the courage to apply, and he was accepted. Had he not spoken English, he would probably have ended up down a coal mine, rather than nurturing a dream to study physics at university.

The British were looking for prisoners of war with no real profession since they seemed to be most useful for work in the German coal mines. "As a young person with a Maturity examination, but no real profession, I was an excellent candidate for coal mines but I didn't feel like joining that profession," said Herwig. Once all the prisoners of war had been sorted and released or sent for trial, Herwig found himself in the employ of the British Army, working as an interpreter for a young reserve officer, a captain a little bit older than himself, and they soon became friends. In the chaos following the war, Herwig had lost touch with his family, who had been expelled from Czechoslovakia to where he did not know, and it would be some years before he found out. For the time being, he was alone in an unfamiliar and devastated land, so when the British officer asked him what he wanted to do with his life, Herwig told him that all he wanted to do was get to a university town and study physics. The British military administration had set up a local headquarters in

Hamburg, and Herwig's officer frequently had cause to visit. "I often have business at the military government in Hamburg," he said to Herwig one day, "why don't you come with me?."

At this point in Herwig's life, chance played an important part in determining his destiny. Had he not been given marching orders to Schleswig–Holstein, he could well have been a Russian prisoner of war. Had he not learned to speak English at school he may well have been down a coal mine. And had he not struck up a friendship with a fellow inmate while a British prisoner of war, he would have had no address to turn to in Hamburg. As it was, one of Herwig's fellow prisoners of war had served in Hamburg during the war. "I met a family in Hamburg by chance when I was a soldier," he said to Herwig, "and they told me that if I ever found myself in trouble they would take me in." When Herwig set off for Hamburg, he set off with their address in his pocket. "The British captain dropped me somewhere near the S-Bahn, and I went to this address at Farmsen, Hasenböge, which was half an hour outside Hamburg by subway and half an hour's walk from the end of the line. When I got there, I rang the bell, and a very nice lady called Frau Palm opened the door." Herwig explained that he wanted to study at the University of Hamburg, and that he needed somewhere to stay for a night while he looked for work and made an approach to the university. "After some hesitation, she said yes, I could stay for the night," recalled Herwig. "In the end this one night became almost four years."

In His Own Words: A Visit to Sanssouci

"I remember my last night in Berlin before leaving for Schleswig-Holstein as if it were yesterday. I'd been in Berlin for half a year, but I'd never found time to visit Sanssouci at Potsdam, and it's a place I really wanted to see. Berlin at the time felt like the end of the world, and I had no idea whether I'd ever get a chance to go back, so I said to my driver, 'take me to Potsdam before we leave, I want to see Sanssouci.' It was late in the evening, and my driver gave me a quizzical look, but we went there anyway, and we arrived just before 10 p.m., when the nightly bombing raids began. It may seem strange to say this, but there was a kind of beauty in the devastation, mixed with the poignancy of seeing this palace of enlightened thinking against the backdrop of Berlin in ruins. The bombing always started in the same way, at ten o'clock sharp, the bombers would drop what we called Christmas trees because of their shape—flares that would light up the skies allowing the bombers to find their targets. They produced a very dramatic illumination of the town, which was all the more strange since they were in the form of Christmas trees, symbols more usually associated with peace on Earth. The illusion was soon broken as the bombs followed the flares, but for just a few minutes by the light of these Christmas trees, I visited the historic castle of Frederick the Great where he had played the flute and discussed philosophical problems with Voltaire and other great minds of the time. I was convinced that I would never come back in my life. Little did I know that my life's journey would lead me to Geneva, where I could visit Voltaire's home

Les Délices, and his Chateau just across the border in the French town that bears his name, Ferney-Voltaire. And I've had the pleasure of going back to Sanssouci on many occasions."

References

1. Merkers Adventure Mines, https://en.wikipedia.org/wiki/Merkers_Adventure_Mines
2. Leopoldina, https://www.leopoldina.org/en/leopoldina-home/

Chapter 3
Studies in Hamburg 1945–1954

In September 1945, Herwig began his studies at the University of Hamburg. Frau Palm had offered him lodgings in the suburb of Hamburg-Farmsen, and the first thing he did once he'd secured somewhere to stay was to apply for a place to study physics and mathematics. He still had his job as an interpreter for the British Military Government, so he had a source of income. He'd carried a copy of his Maturity certificate with him throughout the war, and his excellent grades were good enough to secure him a place.

He'd been lucky to apply as early as he did. The university was in ruins, and soon after Herwig was accepted, the university imposed two conditions on applicants. First, they had to have been living in Hamburg before the war, and second, they had to spend six months clearing up the of ruins of the campus before they could start their academic work.

These conditions did not apply to the early applicants, but the conditions for learning were nevertheless difficult. With a full-time job, he had to be selective about which lectures to attend and which to skip. Those he did attend were often crowded, and students would be sitting on the steps of the lecture halls, or peering around doors to see what the lecturer was saying. One in particular had a peculiar price of entry to his classes. "The lecture halls were not heated," Herwig recalls, "so Lenz always wore a scarf, and he put up a sack at the entrance to the lecture hall where every student had to put a piece of coal, or any other heating material, to help heat his living quarters."

The students benefited from this too. Tutorials were given in Lenz's apartment, in a high-ceilinged and draughty room in which the professor had constructed a small cardboard cubicle around a small coal-fired stove. "When I was examined by him, I had to sit in this little cubicle, sweating next to the fire—from the heat as well as his questions."

The Lenz in question was none other than Wilhelm Lenz, who had been a student of, and then an assistant to, the great Arnold Sommerfeld before becoming director of the Institute of Theoretical Physics in Hamburg. Among his assistants in

© The Author(s) 2024
H. Schopper and J. Gillies, *Herwig Schopper*, Springer Biographies,
https://doi.org/10.1007/978-3-031-51042-7_3

Fig. 3.1 Herwig in his first civilian suit after the war. Photograph taken in the late 1940s in Hamburg (Herwig Schopper's personal collection. ©Herwig Schopper, All rights reserved)

Fig. 3.2 Herwig as a student at the University of Hamburg in around 1948 (Herwig Schopper's personal collection. ©Herwig Schopper, All rights reserved)

Hamburg was Wolfgang Pauli, and Lenz maintained strong links with Niels Bohr in Copenhagen, Max Born in Göttingen, as well as his old supervisor, Sommerfeld, in Munich.

Ernst Ising was Lenz's student. The famous Ising model of ferromagnetism was in fact an idea of Lenz's, given to his student as a problem. "Ising simply carried out the computation that Lenz asked him to do," said Herwig, "so it should really be called the Lenz–Ising model." Lenz remains to this day one of Herwig's favourite teachers, but that's an appreciation that only developed over time. Lenz was not a loquacious teacher, but rather an assiduous user of the blackboard. "I only realised much later how fantastically thought out his lectures were," said Herwig. "He really concentrated on the fundamental aspects of physics, and even today if I have a question on fundamental physics, I look at my notes from his lectures. Probably 80% of my knowledge of theoretical physics goes back to the lectures of Lenz, and I only came to realise this much later." There's a lesson in this, Herwig believes, for those who advocate the kind of rapid and immediate evaluation of teachers that's fashionable today. "I believe that one should not give too much weight to the opinion of students—they understand only much later from which lectures they benefitted most."

Herwig began his studies full of confidence, but was soon brought down to reality. Mathematics at university turned out to be a very different proposition to mathematics at the gymnasium in Landskron, so although he'd mastered the rudiments of calculus at school, at university, he found himself immersed in a whole new vocabulary running from n-dimensional vector analysis to epsilontic reasoning. "I thought the first semester would be rather easy, but to my great surprise, after the first few lectures, I realised I had not understood a single word," he admitted. "It took me quite some time to get used to the scientific language and thinking at the University." But get used to it he did, often working late into the night in the Palm family's kitchen table in Hamburg-Farmsen.

In experimental physics, the first semester of the 1945–1946 academic year was to have been taught by Peter Paul Koch. A student of Wilhelm Röntgen, Koch became a professor at the University of Hamburg when it was founded in 1919. During the war, Koch had been an active National Socialist, and although released by the British before the semester started, he chose to take his own life in October 1945, leaving a vacancy to be filled.

The university recruited a retired professor from East Germany, In reality, however, the lectures on experimental physics were delivered by a technician with no formal physics education but years of experience in preparing demonstration experiments. "He was well able to give the lectures," recalled Herwig, "and they were very popular." The lecture hall was built for around a hundred students, but every day around double that number tried to get in. "Since I was still working for the Military Government, I often came too late to get in. Soon I gave up and did not go to the lectures at all—I just made sure I got the signature of the Professor at the end of each semester as proof that I'd attended the course."

Things took a decided upturn for Herwig with the arrival of Rudolf Fleischmann, a physicist with an excellent track record in research. Born in Erlangen in 1903, Fleischmann worked on isotope separation in Strasbourg during the war. Arrested by the Americans in 1944 because of the relevance of his work to nuclear energy, he spent the last months of the war as a prisoner in the US. Given the benefit of the doubt concerning his political convictions, he returned to Germany a free man in 1946 and went on to have a glittering career, both as teacher and as researcher, first at Hamburg, and then in his native Erlangen. In 1957, he became publicly known as a signatory of the Göttingen Manifesto, which argued against arming the German military with tactical nuclear weapons.

For Herwig, Fleischmann soon became a role model as someone whose curiosity was not limited to a single branch of physics. He'd started his career as a student of the nuclear physicist Walter Bothe, who later won the Nobel Prize. From there, he'd gone on to become an assistant to Robert Pohl, a leading solid-state physicist in Göttingen. "Because he'd worked with both Bothe and Pohl, Fleischmann was not specialised in one area or another," explained Herwig. "He'd acquired and retained a wide understanding, and this impressed me very much. I tried to follow his example all my life—it's really the unity of physics that makes it beautiful."

As well as being appointed to the university's Chair of Experimental Physics, Fleischmann also became director of the *Physikalische Staatsinstitut*, the state physical institute. Initially, nuclear physics was not allowed in Germany, and Fleischmann had to draw on the broad scientific culture that so inspired Herwig Schopper, but as alliances evolved and the Cold War began, restrictions eased and he was able to establish the university as a centre for nuclear physics.

In the early days of his tenure at Hamburg, however, Fleischmann had other things to occupy him. Along with the directorship of the state physical institute came a responsibility to advise the authorities of the State of Hamburg on questions related to physics, such as weights and measures, which are vital to the livelihood of a trading city. After the intensive bombing of German cities at the end of the war, the role took on the altogether more pressing task of bomb disposal. During Operation Gomorrah in 1943, Hamburg had received some 9000 tons of bombs, and many of them were still there, continuing to threaten the lives of the city's remaining inhabitants until long after the war was over. Herwig and his fellow students found themselves devising ways of safely seeking out unexploded ordnance and making it safe.

Herwig Schopper first came to the attention of Rudolf Fleischmann when he registered for his pre-diploma examination, which students sat after two years of studies. The role of the pre-diploma was to select those students worthy of continuing towards a career in research, and Herwig passed with flying colours. "I passed the exam in the fall of 1947, apparently with excellent grades, since Fleischmann immediately offered me a job as a part time auxiliary assistant with the task of preparing experiments for the lectures and taking care of students in practical classes." When Fleischmann later secured sufficient funding to build a new lecture hall, Herwig had an important role to play. "It was really revolutionary," he explained. "The emphasis was given to the possibility to demonstrate experiments and not restrict physics teaching to explanations in chalk on the blackboard. I helped to design the layout of the lecture hall and became responsible for the demonstration of experiments. I spent many evenings trying to get a demonstration to work for a lecture the next morning."

Fleischmann's hands-on approach had been acquired during his time at Göttingen, where he had been introduced to the new teaching style of Robert Pohl, which became famous at the time and changed the style of experimental physics lectures in Germany. Pohl believed that physics should not only be taught on a blackboard, but that the main phenomena should be demonstrated to the students through experiments. "The experiments were sometimes very sophisticated and their preparation took a long time, because we had to be sure that they would work next day, otherwise the students would just have fun in criticising the failed experiments rather than learning anything." Fleischmann also considered it important for students to become familiar with some phenomena even if the detailed theoretical understanding was beyond an introductory course. "This became my first employment in science, and though it was only part time with a ridiculously low salary, it was least enough that I could give up my job at the Military Government and devote my time to physics."

A Diploma in Optics

Fleischmann assigned Fritz Goos, who had been working on optical spectroscopy at Hamburg since the early days of the university, to be Herwig's tutor. Goos was a well-known name in the field of optics, thanks largely to work undertaken during the war years with his student, Hilda Hänchen. The Goos–Hänchen effect is a curious phenomenon linked to the total internal reflection of light at a surface between a dense and a less-dense medium. At a certain angle of incidence, all the light is reflected and nothing gets through. What Goos and Hänchen observed is that light that is totally reflected by a dense medium such as glass appears to penetrate a very small distance into the glass before being reflected such that the reflected light is slightly laterally displaced from the point of incidence. This turns out to be due to a phenomenon whereby the incident light becomes a so-called evanescent—literally vanishingly imperceptible—wave that propagates parallel to the interface. The Goos–Hänchen effect was first reported in Hilda Hänchen's 1943 dissertation, and definitively written up in Goos and Hänchen's 1947 paper, *Ein neuer und fundamentaler Versuch zur Totalreflexion*, in the journal *Annalen der Physik*. That their names remain relatively obscure in modern-day physics circles, despite the fact that the Goos–Hänchen effect is important in many aspects of modern optics, including laser-driven particle acceleration, may be down to the fact that following the wars, the language of scientific publication was shifting from German to English. An English version of their 1947 paper exists in the CERN library [1], but it was not translated until 1972. As this effect exists for all electromagnetic waves it later became interesting for applications such as radar.

Working under Fritz Goos meant that Herwig not only had a very rigorous training in experimental physics, but also an excellent mentor in Hilda Hänchen. She was married to one of the assistants at the institute, Albert Lindberg, who was also a physicist, and Herwig struck up a life-long friendship with them both.

"Goos was an experimenter of the old school," recalled Herwig. "The first thing he told me was that you should never put your trust in dark boxes whose behaviour you do not know exactly, but rather you must understand all the experiment's parts." Thus began Herwig's apprenticeship in techniques such as glass blowing, required to construct vacuum equipment, the technical skills required to make sensitive devices such as electrometers to measure electric charge, and techniques required to operate a precision balance to measure very small weights. "This approach would of course be unthinkable today, with the big instruments and collaborations where everyone has to trust in black boxes doing what they are expected to do, and in the expertise of others," said Herwig. "Glass blowing is no longer part of the undergraduate physics curriculum, but this practical training helped me a lot over the following years."

With the equipment available at the university, coupled with limited funds for research and the fact that even teaching nuclear physics in German universities was banned, Herwig chose a topic in optics for his diploma thesis following a proposal from Fleischmann. This would be the qualification to set him up for a career in industry, and is roughly equivalent to a Master's degree today. Rudolf Fleischmann had many ideas—some excellent, some impractical—and to sort them out he often used his assistants. That's how Herwig found himself working on a problem that Fleischmann had been pondering since his time with Pohl in Göttingen, where they'd observed that the structure of thin layers of alkali metals, such as lithium and sodium, is different to that of the bulk material. Pohl had speculated that alkali metals might have two different crystal phases, one of which was manifest in the bulk, the other in thin layers. "My task was to investigate the optical properties of thin alkali metal layers as a function of their thickness to find out whether Pohl was right or wrong."

The institute was well equipped for optical research, but for the kind of precision that Herwig would need, there was one piece of equipment missing: an optical bench to provide the stability required for the kind of precision measurements he would have to make with interferometry. The solution came from a rather surprising direction. "Goos found out that near to the institute there was an old cemetery that had been badly damaged by bombs during the war, and was being closed," explained Herwig. "Large marble gravestones were being sold off cheaply, and Goos got several of these. By putting such slabs of marble on three tennis balls one got something which was almost as stable as a modern optical bench, and it served my needs well. All the work for my diploma thesis and later for my Ph.D. was done on gravestones."

The technique that Herwig was asked to develop involved shining light on layers of alkali metals of varying thicknesses and studying the reflected and transmitted light. Previous work of this kind had considered the relative change of phase between vertically and horizontally polarised light on reflection, but Fleischmann thought that more insight could be gained into the structure of the sample by measuring the reflected and transmitted phase shifts independently for each of the two orthogonal linear polarisations: the shift in the so-called absolute phase. To do this, Herwig had to have a way of comparing his measurements with a reference beam of light that did not impinge on the metal layers. This required setting up a system of two beams, initially the same phase so that a definite interference pattern could be produced between the reflected or transmitted beam and the reference beam, thereby allowing the phase difference between the two to be determined. In order to get clear interference fringes, the intensity of the two beams had to be comparable and this was achieved with a special device. This had not been done before, and developing the techniques to make such measurements was the task assigned to Herwig for his diploma, which he received on 1 March 1949, paving the way to a Ph.D. studentship.

Fig. 3.4 At Hamburg University after the war, equipment had to be improvised. Herwig's optical bench consisted of a gravestone bought from a nearby bombed churchyard, supported by tennis balls (Herwig Schopper's personal collection. ©Herwig Schopper, All rights reserved)

Restrictions on Nuclear Research in Germany Relax

After the war had been over for three years, the restrictions covering nuclear physics in German universities were relaxed. As a result, Fleischmann was able to introduce some teaching on the subject, although research remained out of the question. In 1948, he had hired one of the world's pre-eminent experts on the matter, Erich Bagge, who had been a student of Werner Heisenberg, and was among the German scientists interned along with Heisenberg by the British at Farm Hall near Cambridge after the war, as the allies tried to understand the extent of the German wartime nuclear programme. Otto Hahn was another internee at Farm Hall. In 1938, Hahn, a chemist, working with Fritz Strassmann in Berlin had been the first to observe nuclear fission, although they did not immediately recognise it as such. They reported their observations to Hahn's long-time colleague, Lise Meitner, who as a Jewish physicist had fled Germany for Sweden. She and her nephew Otto Frisch provided the interpretation—when Hahn and Strassmann had bombarded uranium with neutrons, and observed that the lighter element barium had been produced, the process was due to the fission of the heavy uranium nucleus into two lighter nuclei, along with three neutrons and the release of energy. They published this interpretation in the journal *Nature*. Despite the fact that four people had contributed to the work, the effect became associated most strongly with Hahn and Meitner, and it was Otto Hahn alone who was awarded the Nobel Prize for Chemistry in 1944, making Lise Meitner one of the most notable absentees from the list of those to receive one of science's most prestigious awards. Somewhat ironically, Hahn accepted his prize while incarcerated at Farm Hall. Herwig would later learn about these events from none other than Meitner herself, when he spent some time working with her in Stockholm (see Chap. 5, A sojourn in Stockholm with Lise Meitner).

By this time, Herwig was developing an interest in nuclear physics, but by the time he concluded his diploma work and moved on to a Ph.D., it was still impossible to conduct experimental research in the field. "Fleischmann was trying to establish some experimental nuclear physics in the institute," he recalled, "and later he proposed a Van der Graaf accelerator. This came too late for me because when this machine came into operation I had already left Hamburg." Instead, Fleischmann encouraged Herwig to follow his diploma work through to a conclusion and test Pohl's hypothesis that the alkali metals have a different crystal structure in the form of thin films than they do as bulk matter.

"By determining the optical constants, the index of refraction and absorption, as a function of the thickness of the metal layers produced by evaporation it was possible to explain the measurements by assuming that the thin layers are not deposited immediately as bulk metal but produced by condensation in the form of tiny droplets," Herwig explained. "At first little droplets are formed, and these later join together to form bulk metal layers. The droplets are smaller than the wavelength of the light we used, which only sees the average density of the layers, resulting in apparent anomalous optical constants. Today this explanation is fully corroborated by electron microscope images where one can actually see the individual drops." In his

Ph.D. thesis, Herwig Schopper had demonstrated the hypothesis of Robert Pohl to be incorrect, and in doing so, he had developed a range of formulae for transmission and reflection from thin single layers, and for multilayers, valid for all kinds of electromagnetic waves. Herwig's results were published in *Zeitschrift für Physik.* "These formulae became important not only for my own work, but they also became useful in completely different fields. To my surprise I found out many years later that they were rediscovered. Obviously people don't read old publications, in particular if they were published in German, in journals like *Zeitschrift für Physik* or *Annalen der Physik.*" Herwig's work, like that of Goos and Hänchen before him, had suffered from the wartime hollowing out of German science, as the world's scientific communities looked elsewhere for new developments.

The move from diploma to doctorate was accompanied by a promotion from part-time to full-time assistant to Rudolph Fleischmann, which had important consequences for Herwig both in his professional and private life. Although unable to pursue experiments in nuclear physics, he was able to join in conversations with Fleischmann and the growing team of nuclear physicists he was assembling in Hamburg. "Fleischmann's head was full of ideas," remembered Herwig, "as always some were excellent and some were crazy, and it became the job of his colleagues to test them." One such idea stemmed from Fleischmann's time studying nuclear reactions with Walther Bothe. Just as he'd pointed Herwig in the direction of exploiting the experimental power of polarised light for his thesis, Fleischmann also thought that polarised particles would provide a more powerful tool for investigating nuclear reactions. This was a new idea at the time, but is ubiquitous in particle physics today. At the time, there was no known method to produce a polarised particle beam. Herwig tucked the idea away in his head for future reference.

In the 1920s, Otto Stern and Walther Gerlach had demonstrated the quantisation of electron spin in one of the landmark experiments from the early years of quantum mechanics, and Fleischmann wondered whether a Stern–Gerlach experiment could be adapted to produce polarised proton beams He asked his assistant Herwig to develop such a system. "It took quite some effort and time to convince myself, and to prove to Fleischmann that a Stern–Gerlach experiment works only for electrically neutral atomic beams and not for charged particles like electrons or protons," said Herwig. The experience nevertheless proved useful to Herwig later, because it had led him to study developments subsequent to the Stern–Gerlach experiment, and in particular the Nobel Prize-winning work of Isidor Rabi, who invented the atomic molecular beam magnetic resonance method to study atomic spectra. This is the technique that today underpins the powerful medical imaging tool, magnetic resonance imaging (MRI). Back in the early 1950s, it led to Herwig suggesting to Fleischmann that a modified MRI technique, rather than the simple Stern–Gerlach setup, could be adapted to produce a polarised particle source. Later, when nuclear physics experiments were once again permitted in Germany, Herwig would use this knowledge to build just such a device. "I was intrigued by Fleischmann's ideas," remembered Herwig, "so alongside my doctoral work, I started to build a proton source. It did not provide polarised protons, but the experience came in useful later."

Fig. 3.5 Herwig working on his polarised proton source in Hamburg in the early to mid 1950s (Herwig Schopper's personal collection. ©Herwig Schopper, All rights reserved)

Another of Fleischmann's questions for his team proved to be an important life lesson for Herwig. "I thought classical optics was old-fashioned and I was impatient to complete my thesis and move in the direction of more promising nuclear physics. Only much later did I realise how difficult it is to judge whether a field has become obsolete or not." The question that led Herwig to that conclusion concerned coherent and incoherent reflection of light from alkali metals, and specifically, at what point along the transition from the gaseous start of the metal towards increasing density does the reflected light become coherent. "One of my colleagues, his name was Heinrich Deichsel, was charged with investigating this experimentally. He measured the light reflected from alkali gases in a glass vessel as a function of the gas density. Not much was achieved, but in a way, he was ridiculously close to the idea of 'lasing,' which gives us lasers today. Anyway, I decided to quit optics just a few years before the laser was invented and optics again became a blossoming field. From that experience I learned that one should be very careful when judging a field of physics to be obsolete."

Formative Years

Herwig remembers his days as an assistant to Rudolph Fleischmann in Hamburg as formative for many reasons. In accompanying Fleischmann to conferences, he was able to meet many of the leading figures of twentieth century physics, among them Max Born, Werner Heisenberg, Robert Pohl and Arnold Sommerfeld. Thanks to the efforts to establish the University of Hamburg as a leading scientific and technical university, he also had the chance to rub shoulders with quantum pioneer

Pascual Jordan, and among his Ph.D. examiners was the eminent physical chemist, Paul Harteck, who invented the technique of separating uranium isotopes using centrifuges. Harteck was another of those who had spent some months at Farm Hall after the war. Herwig remembers his Ph.D. examination vividly. "In order to embarrass me, Harteck asked a question that I could not answer. It was only after giving me an excellent grade that he confessed that the answer was only to be found in a paper of his, soon to be published."

In Hamburg, Herwig established many life-long friendships with fellow students and university employees. Among them was Rudolf Kollath, an applied physicist, and later professor at Mainz. "His wife was an excellent violinist and we played together quite often," said Herwig. There was also Albert Lindberg, Hilda Hänchen's husband, who went on to be responsible for the development and production of scientific demonstration equipment for schools at the firm of Linde in Köln, and Hugo Neuert, later a renowned nuclear physicist. Herwig's contemporaries as students included Gerhard Hertz, a great nephew of Heinrich Hertz, who became a professor in Münster, and Reimer Witt who went into industry and made a name for himself at Philips producing television sets.

Herwig received his doctorate on 9 April 1951 and with the exception of a year spent in Stockholm working with Lise Meitner, he remained in Hamburg until 1953. That year, Fleischmann received an attractive offer for a chair at the University of Erlangen. "Since this was his home town," said Herwig, "he accepted. He offered me a post there and I went with him."

In His Own Words: Family Matters

"My years in Hamburg were important not only for the development of my career in physics, but also for personal reasons. When I look back on that time now, I'm struck by how lucky I was, and how easily my career could have taken a completely different turn. It was also in Hamburg that I re-connected with my family who were dispersed around Europe after the war, and that I met my wife Ingeborg, with whom I shared the happiest 60 years of my life.

Before talking about how we met, I'd first like to describe a somewhat strange but important event in my life. After the war, I was a displaced person since my family had been expelled from Czechoslovakia. I didn't know where they were, and they didn't know where I was. We had no contact, and only rediscovered each other when the postal service between Germany and Italy was re-established. I had relatives in Italy, in particular an aunt, a sister of my mother, and from her I learned where my father and my mother were living. They were both safe and well, in separate towns: my mother in in a little town in Bavaria and my father at Wittenberg, the town where Martin Luther had lived. I told my aunt what had happened to me and what a difficult life I was having immediately after the war. My aunt's husband was a sea captain, who commanded a commercial ship after the war transporting coal from Poland to Italy. Why do I tell you all this? Well, to avoid sailing around the Jutland peninsula, my uncle's ship had to pass through the Kiel canal, which links the Baltic Sea to the North Sea. This canal has many locks, and one of the largest is at Kiel, which has a big harbour. One day, I received a letter from my aunt saying that my uncle could pick me up at the Kiel lock as a stowaway and take me illegally to Italy. Since life in Hamburg at that time was still extremely precarious and difficult, with little food and little hope of making a living, and since I faced an uncertain future in Germany, I considered this offer very seriously and I agreed to go. So, with just a small bag containing my belongings, I set off from Hamburg to Kiel and waited near the lock for the arrival of my uncle's ship. Of course, he had told me when he expected to arrive, but ship navigation was rather uncertain in those times. I waited there but he didn't come. Fast communication was impossible—no mobile phones in those days, and normal telephony was also practically not available.

I waited for several days but he did not arrive. Finally, I learned that because of heavy fog, all shipping had been stopped. Since it was not clear how long the fog would last the arrival of my uncle became very uncertain. I could not stay away from Hamburg too long because I would have lost what security I had with my job with the Military Government, and I could not interrupt my study at the university for too long if I wanted to continue there. So, I decided to give up this fantastic possibility, and go back to my life in Hamburg, whatever it may hold. Imagine what would have happened if there had been no fog! My life would have probably taken a completely different direction. So sometimes life is determined by chance, and there is little we can do about it. With hindsight, I'm very glad for those days of fog.

Let me tell you the second incident that determined the course of my life. During my time at the Military Government, I met a young German lady by the name of Ingeborg Stieler in the elevator. Somehow, there was an immediate spark between us, and we noticed a certain mutual empathy. We met again and I learned that she was also working for the Military Government as an English secretary. After some long walks along the river Elbe and several visits to the Hamburg opera house, which had been seriously damaged by bombs and had started some performances using the former stage as an auditorium, we got to know each other quite well.

After many discussions and some hesitation due to the precarity of our situation, we came to the conclusion that we should marry, which we eventually did, and were married for a long and happy 60 years. However, we agreed that the marriage would have to wait until I had a real job to support a family. This sounds very old-fashioned today, but back then it was the norm. We didn't have to wait too long, since this condition was at least formally fulfilled after I passed the diploma examination in 1949 and was employed as assistant at the institute. However, other problems had to be solved before we could marry. Certainly, we wanted to have the agreement of the father of my future wife. He was in the import–export business like many people from families living in Hamburg for generations. Like most of the rest of the population at this time, he knew what a lawyer or a medical doctor did but he had no idea what the chances of a physicist were, and whether such a person could support a family. After some hesitation, and after he got to know me, he agreed to the marriage. His agreement was also essential for a practical reason. Living accommodation in Hamburg was strictly rationed and controlled. Space was limited to about eight square metres per person in the still half-destroyed town, and to find housing for a newly wed couple was practically impossible. Ingeborg was living with her father and her sister in a modest apartment since their house had been completely destroyed during the war. The apartment had two living rooms, a bedroom, a kitchen and a bathroom. Because of the shortage of accommodation, my father-in-law-to-be had been obliged by the authorities to let one of the rooms, and it was occupied by a young man who, among other things, was training to be an opera singer.

We tried to get approval from the authorities to give notice to the opera singer and make his room available for us. But when we went to see the people at the municipal authorities, the first person we spoke to told us that our marriage was our private matter and had no bearing on the rules concerning the distribution of accommodation. We should share the room with the singer, he told us. This was not a joke! Of course, we were very unhappy and went to see his superior, who turned out to be much more reasonable and understanding and agreed that we could have one room of the apartment. So at least we got a roof over our heads, but life in the small apartment was not so easy. Only several years later did we manage to get a larger living space, more independent, but still with only a kitchen corner and no separate bathroom.

We were married on 14 March 1949 at Hamburg-Eppendorf, just two weeks after I'd received my post as an assistant at the institute, but the first few years of our marriage can't have been easy for Ingeborg. At the university, my job as Rudolph Fleischmann's assistant meant that I got involved in teaching experimental physics by being responsible for the preparation of experiments to be demonstrated during the main lectures. Because of this I sometimes had to stay late in the evening and my wife was waiting for me for dinner in vain.

Sometimes she came to the institute to find me, but in the evening the institute was locked so she would call to me from the street outside. This turned out not to be such a good idea, since next to the institute there was the main court of Hamburg, complete with a prison wing to hold delinquents under trial. When my wife tried to call to me, the prison guards thought she was trying, illegally, to contact a prisoner. If that was not enough, barely a year or so into our marriage, I had an opportunity that I could not refuse to go and work with Lise Meitner for a year in Stockholm. Travel was still very much restricted, and the offer was for me only, so Ingeborg had to stay home while I seized the opportunity (see Chap. 5, A sojourn in Stockholm with Lise Meitner)."

Fig. 3.6 Herwig and
Ingeborg's wedding photo,
taken in Hamburg on 14
March 1949 (Herwig
Schopper's personal
collection. ©Herwig
Schopper, All rights
reserved)

Reference

1. http://cds.cern.ch/record/322991/files/CM-P00100636.pdf

Chapter 4
A University Professor, and Establishing New Institutes

The Years at Erlangen

When Herwig arrived in Erlangen in 1953, it was a town of some 60,000 inhabitants, dominated by the university, and the industrial powerhouse, Siemens. Historically, the town had been a seat of nobility, and the Margrave's castle is still a dominant feature. In 1685, when Louis XIV revoked the Edict of Nantes, the Margrave gave refuge to Huguenots fleeing France, even going so far as to build an entire new quarter, Erlangen Neustadt, to house them. This set the town on a course of growth. Another development that shaped the modern-day town came the following century, with the establishment of the university in 1743. Today, housed in a new modern campus, the Friedrich-Alexander University bears the names of the Margrave who established it, and another who later expanded it, cementing its place in the fabric of the town. Many famous scientists have worked at the university including Georg Simon Ohm, whose name became the unit for electrical resistance, and mathematician Emmy Noether, whose eponymous theorem links symmetry to conservation laws—a tenet that underpins much of modern physics.

Erlangen was one of the few towns in Germany to emerge from the Second World War relatively unscathed, but that didn't mean that accommodation was easy to come by. Herwig had been promised an apartment, but when he arrived, he found that it was not ready, and that when it was, the Schoppers would be sharing it with another family. "This was very embarrassing," Herwig recalled, "because my wife was pregnant and we were expecting the birth of our first child." After a few weeks in a hotel, they moved in to the new apartment along with a colleague who had also moved from Hamburg, Horst Wegener, and his family. "Each family had two rooms and a little kitchen, but the bathroom was shared, so life was still somewhat limited." Nevertheless, the apartment was comfortable, and the families got on well.

© The Author(s) 2024
H. Schopper and J. Gillies, *Herwig Schopper*, Springer Biographies,
https://doi.org/10.1007/978-3-031-51042-7_4

Fig. 4.1 Herwig at the carnival at the physics institute in Erlangen in 1958 (Herwig Schopper's personal collection. ©Herwig Schopper, All rights reserved)

In March 1954, Doris Schopper was born. Herwig and Ingeborg's first child went on to forge a remarkable career in public health, earning a degree in medicine from the University of Geneva and a doctorate in public health from Harvard. She held the presidency of the international council of *Médecins Sans Frontiers* in the 1990s, and went on to become a member of the International Committee of the Red Cross, director of the Geneva Centre for Education and Research in Humanitarian Action and a professor in the faculty of medicine at the University of Geneva. In the 1950s, that was all to come, and Herwig carried her in a little seat attached to his bicycle as they explored the surroundings of Erlangen together.

"At the weekends we would explore the beautiful Bavarian countryside around Erlangen by bicycle," Herwig recalled. "A little basket was attached to the handlebars for Doris to sit in comfortably, and there was another basket at the back to carry food for the excursion. In this way, we could extend our range of visits and see many of the cultural monuments in the surroundings." After a few years there was enough money for a car, and the Schoppers bought a Volkswagen. Herwig had passed his driving test when he was serving in the Signals Corps, but he had not converted his military licence into a civilian one. "I never thought I would own a car in my life, so I had to start driving lessons all over again."

In Erlangen, the young family thrived and Herwig rekindled his love of music by installing a piano in one of the two rooms "But there was a problem. We lived on the ground floor, and above us was a mathematician who was also the dean of the faculty, so in a way, he was my boss. He could not tolerate noise, so I could only play the piano when he went out. My wife would look out of the window and tell me to stop when she saw him coming."

After the war, the university flourished, and by the time Herwig arrived, its reputation in physics was growing. "There was an experimental physics institute which had become quite well known thanks to the work of Bernhard Gudden," explained Herwig. "Fleischmann had received his doctorate under Gudden between the wars, and it was when Gudden accepted a chair in Prague that he moved to Erlangen. Gudden was an expert in superconductivity, which was one of the most interesting issues of the time, so the institute had good infrastructure for low temperature experiments."

Superconductivity, however, was not the main research interest of Fleischmann's new group, and Herwig found himself taking his first steps into science administration in parallel with his research. Many people had come with Fleischmann from Hamburg but Herwig was the most senior assistant. "I arrived in Erlangen with the main objective of becoming a Dozent," he explained. "I would translate that as lecturer, but it's a bit more than that. As a Dozent, you give lectures, of course, but although you don't have voting rights, you are considered to be a member of the faculty, so it's an important step in an academic career." Herwig soon fulfilled all the conditions to achieve such a promotion, and as Fleischmann's Dozent, he had the additional task of converting the university's physics research infrastructure to match the needs of the new group.

Measuring the Circular Polarisation of Gamma Rays

In Erlangen, Herwig pursued two lines of research. "I continued the beta decay work that I'd started with Lise Meitner in Stockholm, in particular I started to look at beta gamma correlations, and I also continued the work I'd begun in Hamburg on a polarised proton beam source."

Sometimes when an atom undergoes beta decay, emitting an electron, the daughter nucleus rapidly emits a gamma ray photon. When this happens, the angular correlation between the beta particles and the gamma ray photons can be used to glean information about the structure of the nucleus that has decayed. This was an established technique by the 1950s, but Herwig wanted to take it a step further by measuring the circular polarisation of the gamma ray photons, which had not been done before. Techniques existed for measuring the circular polarisation of visible light, but not for photons with the much higher energy of gamma rays. Herwig had read a paper by Dutch theoretical physicist, Hendrik Anton Tolhoek, suggesting that it could be done by scattering the photons from magnetised iron, in which the electrons are aligned. While at Erlangen, Herwig established and perfected the technique, publishing several papers shedding light on the structure of beta-emitting nuclei. It was a technique that would serve him well later, when he was a visiting scientist at Cambridge University looking at parity violation in beta decay.

The First Spin-Polarised Proton Beam Source

As a Dozent, Herwig now had a doctoral student of his own, Günther Clausnizer, who would go on to become a professor at Justus Liebig University in Giessen. Together with Fleischmann, the two of them pursued the work Herwig had initiated in Hamburg, building the world's first polarised proton source, and publishing a paper about it in August 1956 in the German language journal, *Zeitschrift für Physik* entitled *Erzeugung eines Wasserstoffatomstrahles mit gleichgerichteten Kernspins* (*Generation of a hydrogen atom beam with aligned nuclear spins*). At the time it was common practice that the director of the institute co-signed most publications. A significant achievement though it was, the reach of German-language journals was not what it once was, and the paper was little read, leaving polarised proton sources to be re-invented at a later date, and in another place.

Meeting Other Scientists

Erlangen proved to be a stimulating environment for a young research physicist. As well as the experimental group headed by Fleischmann, there was an institute of theoretical physics where Herwig met an assistant of about his age, Hermann Haken, who later founded the discipline of synergetics. There was also an institute of applied physics, with Erich Mollwo and his assistant Gerhard Heiland. With Heiland and Haken being at the same point in their careers as Herwig, the three became firm friends, a relationship that endured long after Herwig's time in Erlangen.

Industrial research was also very important in Erlangen at the time, with Siemens having a major presence in the town. The company not only had production facilities there, it also ran a research laboratory that had links to the university. Among the scientists that Herwig got to know at Siemens was semiconductor pioneer Heinrich Welker, widely acknowledged as the inventor of type III–V semiconductors, which he developed in Erlangen and are ubiquitous in the electronics industry today. Earlier in his career, while working for Westinghouse in Paris, Welker, along with Herbert Mataré, narrowly lost out on being credited with the invention of the transistor. Welker went on to become director of all of Siemens' research laboratories, and Herwig recalls one promotion with amusement. "One of the perks of the new rank that Welker had been promoted to was that the doormat in front of his office would be flush to the floor. The works were duly undertaken, but then the promotion was delayed for some administrative reason. The mat had to be raised up, only to be re-sunk into the floor when the appointment was confirmed. There was some silly bureaucracy at Siemens, but in general it was very inspiring to have the Siemens lab close to hand." Another interesting personality who Herwig met was Wolfgang Finkelnburg, a well-known physicist with whom Herwig later had many contacts concerning the popularisation of physics.

Moving to Mainz and the Foundation of MAMI

Herwig's stay in Erlangen was a short one. In German physics circles, he was starting to make a name for himself, and in 1958 at the tender age of 34, he was offered a Chair of Nuclear Physics at the University of Mainz. "Fleischmann wanted to keep me in Erlangen, and told me he would do everything he could to get me a chair there," Herwig recalled. But his mind was made up. "There are two ways to make an academic career—either you stay in an institute under the umbrella of a famous and powerful personality, and eventually move up to a chair, or you move on and become independent. I did not want to stay rooted in one branch of research, so much to Fleischmann's dismay, I accepted the offer in Mainz."

When the Schoppers arrived, they found a very different town to the one they'd been living in. Unlike Erlangen, Mainz had been heavily bombed during the war, and some 80% of its historic centre had been destroyed. In the 1950s, archaeologists were still excavating the bomb sites to discover the historic treasures of the city, which dates back to Roman times. The university itself, although having ancient roots, had languished, and was only re-established in 1946, under the French post-war occupancy of the Rhineland Palatinate region. Even though the population of the town had fallen sharply during the war years, accommodation was scarce, and the Schoppers found themselves involved in a long search for somewhere to live. Young Doris, by now four years old, would play a decisive role. "They had promised to provide housing, but there was nothing there. While I was at work, my wife went from one agent to another with my daughter, who eventually decided she'd had enough. She crawled under the agent's desk and said: 'mother, I will not move until you've found somewhere for us to live!' Finally we found two rooms under the roof of a little house in the suburbs, a place called Mainz-Gonsenheim. It was a nice part of town, but still it was very modest. One of my former school friends came to visit, and I remember him exclaiming: 'My God! I thought professors would live differently—what a shabby apartment.'" It was not long, however, before the university lived up to its word, and the Schoppers were able to move into newly built accommodation for university personnel in Göttelmannstrasse. They moved there in 1958, well in time for their second child, a son, to make his entrance.

Andreas Schopper was born in October 1959. "Andreas had a rather eventful childhood, being born in Mainz, but then spending less than a year there before moving to the US, and later coming with me to Karlsruhe, Geneva and Hamburg," explained Herwig. "He chose to study physics at Basel although I never once tried to influence him, knowing that a career in physics can only be successful if driven by great enthusiasm for the field." Perhaps some of his father's own enthusiasm rubbed off on the young Andreas, growing up while Herwig was at DESY and CERN. "Eventually Andreas got a permanent position at CERN after I retired," continued Herwig. "He worked at the storage ring LEAR and later became a member of the big LHCb experiment at the LHC where he accepted several important tasks in the collaboration. He was also elected president of the Swiss Physical Society in 2012, a rare appointment for a non-Swiss citizen."

Geretsried Sommer 66

Fig. 4.2 The Schopper family at Geretsried, south of Munich, in the summer of 1966. Left to right, Doris, Herwig's mother Margarethe, Ingeborg, Andreas and Herwig. (Herwig Schopper's personal collection. ©Herwig Schopper, All rights reserved)

Mainz University's physics department was located in a former army barracks that had survived the war, and Herwig's job was to establish a new institute for research in experimental nuclear physics, which by this time was allowed in West Germany. There was no nuclear physics at Mainz when he arrived, but by the time he left just two years later, the university would be on the way to hosting an important centre for fundamental research in the field. Although the university's physics department was modest at the time, Mainz was also home to a Max Planck Institute (MPI) that specialised in radiochemistry.

Founded in 1911 as the Kaiser Wilhelm Society, the Max Planck Society got its current name in honour of a former president in 1948, and is a state-funded association of research institutes. "The Max Planck Society's policy was, it has changed now, but it was to create an institute around a famous personality where they could do what they wanted. When they died, the institute would be dissolved, or converted to another subject." The MPI for nuclear radiochemistry in Mainz was built around one of the most eminent scientists of the day, Josef Mattauch.

The turmoil imposed on central European science by National Socialism had not bypassed Mattauch, whose career it had very much influenced. After a promising start at the University of Vienna, a Rockefeller Fellowship gave Mattauch the opportunity to work at Caltech in 1927–1928 on the development of mass spectrometry. Returning to Vienna he continued to develop the technique, rising rapidly through the ranks, eventually succeeding Lise Meitner as head of the university's department of mass spectrometry when she left Austria to go to Berlin. "Mattauch was famous because he had been one of the first to build a powerful magnetic mass spectrometer to analyse the nuclei produced in nuclear reactions," recalled Herwig. "By the time I got to Mainz, he was head of the MPI. One of the department heads was Fritz Strassmann, who had worked with Otto Hahn and Lise Meitner in Berlin. In the late 40s, they enthusiastically supported an initiative to make Lise Meitner head of the University's physics department, but she turned the offer down."

Among the staff at the MPI was Swiss nuclear physicist, Hermann Wäffler. "He had some experience with accelerator physics, and I thought that together we might build up something there. There was a lot of bureaucracy, and an unforeseen hurdle to overcome in the form of a senior member of the university's management. I had to get authorisation and money to construct a building for the university institute, and I agreed with Wäffler that we would make an application for an electron linear accelerator of about 100 MeV or something like that. We prepared everything, and the first thing was to get approval from the university. There was a lot of resistance, and for a long time I couldn't work out why. It was not a matter of money, because funds would have to come from the Federal Ministry of Research, in Bonn." Sometimes it is more difficult to find out why there is resistance and where it comes from than it is to solve the problem itself, and that proved to be the case in this instance. The senior management officer of Mainz University was known as the *Kurator*, and when Herwig found out that the *Kurator* too was waiting for university accommodation, things started to fall into place. The new apartment building the *Kurator* was slated to move into was right next to Herwig's proposed accelerator laboratory, and that made him nervous. "Not being a physicist, he thought there might be a risk, so I told him that I'd be prepared to move in as his neighbour. Although this never happened, I got the university's approval. We submitted the proposal to the research ministry, and after going through various committees, the outcome was positive and a linear accelerator was built."

Today, the laboratory that Herwig Schopper and Hermann Wäffler established in Mainz has evolved to become the Mainz Microtron Laboratory (MAMI), core of the university's nuclear physics department and among the largest university-based accelerator facilities in Europe. Herwig, however, did not wait for the linear accelerator he'd commissioned to see the light of day. His restlessness, combined with his still-growing reputation as a physicist with a strong managerial and administrative bent, meant that he had no shortage of offers, and two of them proved too good to refuse. "I'm a restless man," he confessed. "I didn't stay in Mainz. It was my successor, Professor Ehrenberg, who put the linac into operation and developed nuclear physics in Mainz." Herwig was heading to Karlsruhe, with a little detour via Ithaca, in New York state.

The Foundation of CERN and DESY Leads to Difficult Decisions

While Herwig had been moving around Germany, there had been an important development in fundamental nuclear research in Europe with the establishment in 1954 of the European Organization for Nuclear Research, CERN, just outside Geneva, Switzerland. With 12 founding member states, CERN was the brainchild of a number of visionary scientists and diplomats who saw fundamental science as the glue that could stick a war-torn continent back together. The idea for a laboratory like CERN was first put forward to the United Nations by the French delegation as early as 1946, and as the idea matured and evolved it gained support from across the continent and further afield.

The scientists recognised that European countries could only become competitive again by joining forces, while the diplomats saw science as a neutral language to

Fig. 4.3 Wolfgang Gentner (left) architect of CERN's first accelerator, the Synchrocyclotron, became chair of the Laboratory's Scientific Policy Committee in 1968. Here he is in discussion with German delegate Wolfgang Paul. Paul went on to win the Nobel Prize for physics in 1989 (©CERN, All rights reserved)

promote dialogue between nations. "In 1949, the Swiss diplomat and writer, Denis de Rougemont, organised a European cultural conference in Lausanne, bringing together diplomats from countries including the UK, France and Germany," explained Herwig. "At that meeting a message was delivered from French Nobel Prize-winning physicist, Louis de Broglie, advocating a European laboratory." It was then that the idea really took hold, and in 1950 at the UNESCO general conference in Florence, an American, Isidor Rabi, tabled the motion that would lead to the establishment of CERN. By the end of the 1950s, CERN had established two important milestones—in 1957, it had brought into operation the highest energy particle accelerator in Europe, a 600 MeV synchrocyclotron (SC), and in 1959, the laboratory commissioned the highest energy particle accelerator in the world, the 28 GeV proton synchrotron (PS). Although the PS would not hold that accolade for long—the American alternating gradient synchrotron (AGS) at Brookhaven on Long Island would soon surpass it— European fundamental physics research was back on the map.

The creation of CERN transformed fundamental physics research in Europe. The larger, more wealthy member states were able to maintain domestic facilities for fundamental physics in parallel to CERN, and over the years, a policy evolved of developing national or regional laboratories that would be complementary to the big European laboratory in Geneva. By the early 1960s, with the SC and the PS in routine operation, and the tradition of competitive competition between CERN and the US

Fig. 4.4 Three of CERN's founding fathers, left to right: Pierre Auger, Edoardo Amaldi and Lew Kowarski pictured in 1952 (©CERN, All rights reserved)

Fig. 4.5 Auger and Amaldi were together again to celebrate the Laboratory's 30th anniversary in September 1984. Behind and between them is Denis de Rougemont, who also played a key role in the establishment of CERN (©CERN, All rights reserved)

labs firmly established via the PS and AGS, thoughts were turning to the next big machines. In the USSR, a 70 GeV machine was already under construction, while in the US, design work had begun on a 200 GeV machine, with talk of energies as high as 1000 GeV.

CERN's Director-General, Viki Weisskopf, and the chair of the lab's Scientific Policy Committee, Cecil Powell, convened a meeting of leading European physicists on 7 January 1963 to thrash out European plans. They agreed that a wider group should be constituted to consider the future of accelerators in Europe. That group met on 17–18 January 1963, and constituted itself as the European Committee for Future Accelerators (ECFA). Among the German delegation was a certain H. Schopper. By this time, the policy of regional labs had been formalised into a 'summit programme' at CERN, which would be built around a major facility that would require the efforts of all CERN's member states, and a 'base of pyramid' programme with national and regional labs hosting more modest facilities that would complement the big machines at CERN. An ECFA working party noted that member states that had strong domestic programmes were able to benefit more from their membership of CERN, and flagged up the kinds of facilities that might constitute the base of the pyramid.

Fig. 4.6 Viki Weisskopf was joined by some equally distinguished physicists at a colloquium held in his honour on the occasion of his 80th birthday. Left to right: Val Telegdi, Leon Lederman, Viki Weisskopf, Antonino Zichichi, Herwig Schopper (©CERN, All rights reserved)

This was the context surrounding Herwig's move from Mainz, and it proved to be decisive in the direction his career would take, largely due to the influence of Willibald Jentschke, an Austrian physicist who had moved to the US after the war and become head of the cyclotron laboratory at the University of Illinois. It had taken the University of Hamburg some time to replace Rudolph Fleischmann when he left for Erlangen, but Jentschke was the person that the university eventually settled on. He proved to be somewhat hard to get, laying down conditions before accepting the post and moving to Hamburg in 1956.

Jentschke originally wanted to establish a base of the pyramid accelerator laboratory at the university in Hamburg. He had gained experience with electron accelerators during his time in the States, and as CERN was a laboratory built around proton machines, his condition for moving was an undertaking from the university that it would support him in a bid to establish an electron accelerator-based facility for high-energy physics, as the emerging field was beginning to be called, in Hamburg. "The authorities in Hamburg agreed to provide, I think it was about 5 million deutschmarks for the accelerator," said Herwig, "and give support for an application for funding from the Federal Ministry in Bonn. What is five million? Not much for a high-energy physics facility, even at that time." Jentschke started discussions with influential physicists across Germany—people such as Wolfgang Gentner in Heidelberg, Wolfgang Paul in Bonn, and Wilhelm Walcher in Marburg. "Coming from the States," explained Herwig, "he was ambitious. He saw an opportunity for something more than just a university laboratory—he wanted to create a national laboratory for high-energy physics in Germany at the base of the ECFA pyramid." Jentschke secured undertakings from Bonn and the state of Hamburg to fund the facility, and on 18 December 1959, the *Deutsches Elektronen-Synchrotron* (DESY) was established on an old airfield in the Hamburg suburb of Bahrenfeld.

Jentschke chaired the DESY directorate until 1971, but in the early days, his problem was not just establishing the lab, he also had to build up a community to use it. "There weren't many high-energy physicists in Germany," explained Herwig. "He hired some of his former colleagues from the States, among them Martin Teucher and Peter Stähelin, who played decisive roles in setting up an experimental programme at DESY." But even with experienced people such as Teucher and Stähelin at DESY, Jentschke still needed to build up a user community in Germany. "So he came to me," said Herwig, "and asked if I'd be prepared to go to the States to learn how to use a high-energy electron accelerator, and set up a user group at Karlsruhe when I got back." The place to be for circular electron machines at the end of the 1950s was Bob Wilson's laboratory at Cornell University in upstate New York, and Jentschke offered to arrange a year's placement for Herwig there.

This left Herwig with some negotiating to do at Karlsruhe. The university had offered him a professorship, and in addition a contract as head of a new Institute at the *Kernforschungszentrum* (KfK) which had a small cyclotron of about 60 MeV used for radiochemistry. The university and the KfK were two distinct organisations but worked closely together. Herwig was keen to accept this promising offer, but he did not want to miss the opportunity of a year at Cornell. "I said to Karlsruhe that I'd like to come, but under the condition that I formally start my job with a one-year leave of unpaid absence to go to the United States." They agreed, and in the summer of 1960, the Schopper family set off for Ithaca. A year later, Herwig took up his position in Karlsruhe. "The University of Karlsruhe had a great tradition in physics. It was there that Heinrich Hertz had discovered electromagnetic waves, for example, and Wolfgang Gaede had developed vacuum pumps. Recently, in a building on the outskirts of the town the university had installed a small accelerator, a betatron, for nuclear and solid-state physics."

Fig. 4.7 The Karlsruhe *Institut für Experimentelle Kernphysik* in 1964 (reproduced courtesy of the KIT archives, 28028/04181. ©KIT, All rights reserved)

Herwig had another condition for accepting the position at Karlsruhe. "My second condition for accepting the university's offer was that in addition to being a full professor at the university, they would make me director of the two individual institutes, united under the same name as the Institute for Experimental Nuclear Physics." Herwig's condition was accepted by both organisations, and *the Institut für Experimentelle Kernphysik* (IEKP) was established with Herwig at the helm. "Although the financing and administration were quite different, in daily life it seemed to be just one institute," recalled Herwig with satisfaction.

"I thought that by combining the two institutes, we would have the advantages of both: as a university professor, one has great independence with respect to the research you do, and you have close contact with students, which is a great pleasure as well as allowing you to attract the best to the institute. At the research centre, you would have good infrastructure, but less freedom: the research policy would be determined by government, and they wanted more applied research. Combining the two would give me the freedom to do the research I wanted, with the resources that come with a national research centre." Herwig wanted to ensure that his new position would be right for him, but he also had a bigger picture in mind. "I thought that the stability of both organisations would be improved in the long run by bringing KfK and the university together," he explained. "In that attempt I was supported by Erwin Becker, director of the Institute of Applied Physics and Walter Seelmann-Eggebert, a former student of Otto Hahn and director of the Institute of Radio Chemistry, both friends of mine and professors at the university." It was a visionary dream, but it took much longer to realise than any of them could have imagined. "After difficult discussions over many years my initial dream has finally become reality," said Herwig. "It took until 2009, but today KfK and the university have merged under the name of the Karlsruhe Institute of Technology (KIT), echoing somewhat MIT in the USA."

Herwig's concept for the IEKP looked deceptively simple, but in practice Herwig found himself engaged in a battle with industry for the favour of the ministry in Bonn. As chair of the scientific council of KfK he argued for a multidisciplinary research centre developing reactor technology for potential commercial application as its main activity, with a parallel strand in basic research. "It was very hard to convince the Ministry because of the pressure from German industry, so I failed to realign the research centre with basic and applied research," he recalled. "I'm very sorry about that for two reasons, firstly because I had hoped to create a high-energy physics centre in Karlsruhe, which would provide two legs for KfK to stand on, applied technology and basic research, giving it more stability and independence from fast-changing industrial preferences, and secondly because KfK missed another great opportunity, which was eventually realised in Darmstadt."

High-Energy Accelerators at Karlsruhe?

Karlsruhe is right next to the border with France close to Strasbourg, and Herwig's plan was to establish a regional centre for both countries. "There was already DESY," he explained, "but it was not clear how big DESY would become, and I thought there was room for another base of the pyramid lab in a big country like Germany." This plan never really got off the ground, but another opportunity soon appeared for KfK.

Among the people Herwig had been talking to about establishing base of the pyramid facilities in Germany was Christian Schmelzer, a professor at the University of Heidelberg who was pushing to set up a centre for heavy-ion physics in Germany. Schmelzer had previously been at CERN, where he played a part in getting the lab's big machine, the PS, up and running in 1959. Just as Jentschke had chosen an area that CERN was not involved with—electron accelerators, so Schmelzer had chosen another complementary area in the form of heavy ions. "There was no place for such a facility in Heidelberg, and the university was not interested in hosting it anyway, so together with Schmelzer, we considered establishing it at KfK, but because of the resistance from industry, the Ministry turned us down."

Schmelzer persevered, however, and did eventually get his heavy ion research centre built. In 1969, the *Gesellschaft für Schwerionenforschung* (GSI) opened its doors for the first time in Darmstadt. For Herwig, however, this was a frustrating period.

When he returned from Cornell, Herwig had a promise to fulfil to Willibald Jentschke: to set up a DESY users' group at Karlsruhe. This allowed him to pursue his scientific work with the full support of both KfK and the university. He secured two new chairs at the university, and in 1965, recruited Anselm Citron and Arnold

Schoch from CERN. After Schoch passed away in 1967, low-temperature physicist Werner Heinz stepped into his shoes two years later, and together, Citron, Heinz and Schopper hatched a plan to design and build a superconducting proton synchrotron with an energy of around 100 GeV in Karlsruhe. "By this time, CERN also had new ideas," recalled Herwig. "John Adams was proposing a much bigger machine, a Super Proton Synchrotron (SPS) of 300 GeV, and this was being discussed at the international level. It was becoming clear that the ECFA pyramid might no longer be a possibility for Karlsruhe."

A Second CERN Laboratory and the SPS

CERN's founding convention specifies that the organisation should be responsible for the construction and operation of "one or more international laboratories," and with the SPS being seriously considered, many of CERN's member states had proposed sites for the new CERN facility. Germany was faced with a choice: either support the SPS project, or go ahead with the superconducting facility at Karlsruhe. Both would not be possible.

Eventually, Adams' argument that it made sense to build the SPS at the Geneva site where it would benefit from the existing infrastructure, notably using the PS as an injector, won the day. "Adams had already been appointed Director-General for the new laboratory before the site was decided, so that was the origin of the two Directors-General for CERN," said Herwig. "There was a lot of politics at the time, and it was clear that Adams wanted to become Director-General of the new laboratory. If the SPS had been built as a project of the existing CERN laboratory, it's not clear that he would have become Director-General." As it was, in 1971 Adams became Director-General of a new CERN laboratory, CERN II, which was established in the French village of Prévessin, just across the border from the original CERN site. "That caused problems for me when it was my turn to become Director-General and I had to unify CERN I and CERN II," said Herwig, "but I'll come back to that later." Germany joined CERN's SPS project, and Herwig's last attempt to bring a base of the pyramid facility to Karlsruhe came to nought. "I supported the SPS, and I must admit that I'm a little proud that I could convince my colleagues to abandon the idea at Karlsruhe in favour of the SPS."

Successes in Science

Herwig's first forays into scientific administration might have met with varying degrees of success, but his scientific work thrived in Karlsruhe. "These were perhaps my most fruitful scientific years," he recalled. "I split my work in several ways. I continued with my own nuclear physics work on beta decay with some colleagues who had come with me—I still keep in touch with some of them today. The most senior was Helmut Appel who later became a professor at Karlsruhe. We published important results on internal *bremsstrahlung* following beta decay by electron capture, which showed that parity violation was 100%."

Another strand was a polarised proton source for accelerators. "We tested it with the Karlsruhe cyclotron," explained Herwig. "I must admit I didn't contribute much to that, but my colleagues got it working." Herwig also fulfilled his commitment to Willibald Jentschke by setting up a Karlsruhe user group at DESY, which conducted the first experiment by an outside team, and the first at DESY using counter detectors rather than bubble chambers. "It was more or less a copy of what I'd done at Cornell, elastic electron scattering from protons and neutrons, so my year in Ithaca paid off," he continued. "But at both Cornell and DESY we neglected inelastic electron scattering and thus missed the discovery of 'partons', which were later identified as quarks. This discovery was made at SLAC, perhaps because they had a closer relation between theory and experiment."

Accelerator Technology and Superconducting Cavities

One aspect of his time at Karlsruhe that Herwig is most proud of is the work he initiated there into accelerator technology. "I thought that since KfK was a technical centre, we should do some accelerator technology development. At the time, superconductivity was already being used by industry for medical applications for strong magnetic fields in nuclear magnetic resonance imaging, NMR, which nowadays we call magnetic resonance imaging because people got frightened by the word nuclear, but I'd heard about superconducting cavities to accelerate particles." When Karlsruhe was bidding for a base of the pyramid facility, a superconducting proton machine was what Herwig had in mind, and by the time they gave up on the bid, they had already invested a lot of effort in this direction. "Some tests had been done with high-frequency superconducting lead cavities—at the time, lead was the favoured superconductor," Herwig pointed out. "So we started with lead, but later when it became clear that niobium is much better than lead, we switched, first to niobium-coated copper cavities, and later to pure niobium. We were the first group in Europe to investigate superconducting radiofrequency cavities."

Herwig developed a strong interest in novel accelerator technologies, and although there were some dead ends, the legacy of this work would prove to be significant for future accelerator projects. One curiosity from the annals of the field was a type of machine that rejoiced in the name of the smokatron. "A new idea that had come up and was being developed, above all in the Soviet Union, was the so-called smoke-ring accelerator where one created a ring of electrons in a strong magnetic field. As the ring expanded, the electrons accelerated. The idea was to load these rings with heavy particles like protons that would get accelerated too. They called it a smokatron by analogy to the smoke rings that smokers used to blow. I must admit I wasted a year of work on the smokatron before we realised it would not work. Today, there's a related principle, wakefield acceleration, that looks promising, but back then, the smokatron was a dead-end."

The work on superconducting cavities proved far more successful. "After CERN got the SPS, nobody was interested in superconducting linear accelerators, so we started to think about using transverse fields in superconducting particle separators." To develop such devices, Herwig invited an expert from CERN to spend a year at Karlsruhe applying superconductivity to particle accelerators. "Herbert Lengeler came to us from CERN. He worked with us on superconducting cavities and took that technology back to CERN. This work gave superconductivity an immediate application in separating a mixed particle beam into its components." The technique was taken up not only at CERN but also at the big Soviet laboratory in Protvino. Years later, Lengeler would become an important member of the group that developed superconducting accelerating cavities for the large electron positron collider (LEP), allowing it to accelerate electrons to 104.5 GeV, a record that still stands to this day.

In His Own Words: Who Cares About Neutrons? The Hadron Calorimeter

"In 1964, I was invited to spend a year at CERN. Well, at that time high, high-energy physics with electrons and protons were completely separate worlds. We hardly spoke to each other. There were separate conferences. There was an international high-energy physics conference for electrons and neutrinos, and a separate one for protons with hardly any communication between the two. One dealt mainly with weak interactions, and the other with the strong interactions. CERN was a proton lab. I arrived there with the stamp of electron high-energy physics, and I was an outsider. So what could I do at CERN? Either I could join an existing group or I could propose something myself. At first, I tried to join a group, but that was hard, since most of the big established groups were not keen to accept an outsider like me who might have his own ideas. So I joined a relatively small group that was led by a Norwegian called Arne Lundby. He was very friendly, and he accepted me in to his group, which was studying pion production. That was a start, but after some time I thought, "Okay, I have the group in Karlsruhe behind me, we could propose something new." Karlsruhe did not yet have any relationship with CERN, so I had the idea of establishing a user group from Karlsruhe at CERN. A group led by Kai Runge, a professor at Freiburg im Breisgau, joined us and together we submitted a proposal.

Studying the experimental programme of CERN, I discovered that there was no experiment looking at the production of neutrons. There was no detector for neutrons. So I asked myself, "how can one detect neutrons and analyse them?" Neutrons have no electric charge, and so they cannot be deflected and analysed by magnetic fields like charged particles. At Cornell, I had learned how to observe photons, which are also neutral, by using scintillators, which produce light when hit by charged

Fig. 4.8 Arne Lundby (right), who accepted Herwig into his group, with Kjell Johnsen in November 1974 at an intersection point of the Intersecting Storage Rings, ISR, the world's first hadron collider. Johnsen was ISR project leader ()

particles. By making the photon convert into a shower of charged particles, and then collecting all the light, you get a signal that is proportional to the original energy of the particle. This method to detect photons was invented by Bob Wilson, who called it, somewhat misleadingly, photon calorimetry, although no calorific effects are involved. I thought, "Why not use that method also for neutrons?" When I looked at the literature, I found that thin scintillation counters had been used to detect neutrons in cosmic rays, but not to measure their energy or direction, just the location where a neutron came down. I asked myself, 'Why not use the total absorption to measure the energy of neutrons?' and I decided to make a proposal to build a neutron detector at CERN. All the preliminary work was done with my group at Karlsruhe, where we built the first neutron counter for high-energy physics experiments. We did our first experiment at the proton synchrotron, and later we used this detector at the Intersecting Storage Rings. Finally, the CERN Karlsruhe group made a proposal for the Serpukhov accelerator in the Soviet Union which, at the end of the '60s, was the most powerful proton machine in the world with an energy of 70 GeV.

Another important development at that time was that we optimised this spectrometer for our experiment. It sounds pretty obvious today, but at the time I think we were the first group to use a Monte Carlo computer program to optimise a detector.

Our neutron calorimeter was built as a sandwich of iron plates and scintillator sheets. In order to stop the neutrons you need a heavy material, like iron, and between the plates of iron you put scintillator plates that collect the light produced by the particle shower produced when the neutrons are absorbed. So we knew we wanted a sandwich of iron and scintillator plates about one metre long, but we needed to find the best distribution of iron and scintillator to get the best energy resolution. I learned that at CERN there was a group responsible for radiation safety under the responsibility of Klaus Goebel, whose job was to make sure that none of the radiation produced in the accelerators could escape. For this purpose the Safety Group had developed a Monte Carlo program to perform calculations for the absorption of radiation. I think it was one of the first Monte Carlo programs ever used at CERN, and we were allowed to use it to model the particle cascades produced in our iron absorber.

It was very advanced for its time, even if it looks antiquated now. During my stay at CERN, I learned how to use CERN's IBM computer and to program it in the Fortran language. I was running around with one-metre-long steel boxes containing IBM punch cards to do the calculations. The trouble was, if you made one little Fortran error, you had to wait for next day to find out. It was very cumbersome, I can tell you. Eventually, we used a calorimeter to study the neutron production at the PS and later at the ISR, and we did an experiment at Protvino in the USSR.

With hindsight, that first experience at CERN was very important, but at the time, and much to my dismay, most people laughed at me for my idea of a neutron calorimeter. They said, 'A calorimeter for neutrons? Complete nonsense. Who cares about neutrons? The only spectrometer you need is a magnetic spectrometer where you can get high precision for charged particles.' Years later, with the arrival of

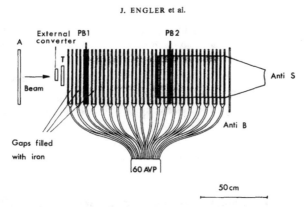

Schematic layout of counter II with 40 scintillators. T is an external trigger counter; PB1 and PB2 are internal trigger counters. The counter A is in anticoincidence for neutrons and in coincidence for protons.

Fig. 4.9 The set-up of the first hadron calorimeter built by the Karlsruhe group at CERN (NIM 106 (1973) 189–200). This calorimeter was also the first instrument at CERN to be computer-optimised. Although unfashionable when Herwig proposed it, the calorimeter he built when he was first at CERN proved to be a very useful tool (© Elsevier, All rights reserved, Nucl. Instr. Methods 106 (1973) 189–200 [1])

Fig. 4.10 Herwig Schopper (right) with CERN Director-General Willibald Jentschke at Protvino in 1972 discussing CERN-Soviet collaboration in physics with their Soviet counterparts (©CERN, All rights reserved)

collider experiments, it turned out that it was not only high resolution that mattered, but that you also need to cover a large solid angle to capture as many of the particles produced as possible. Neutral particles were just as important for the physics, and suddenly calorimeters became fashionable. Nowadays, all collider experiments include large calorimeters that have improved dramatically over the years. I must say, I suffered a lot from being ridiculed by my colleagues, but I got the last laugh."

Reference

1. Engler J et al (1973) A total absorption spectrometer for energy measurements of high-energy particles. Nucl Instr Methods 106:189–200. https://doi.org/10.1016/0029-554X(73)90063-3

Chapter 5
The Travelling Years - Stockholm, Cambridge and Cornell

While Herwig was moving around from place to place in Germany, establishing institutes and moving swiftly upwards through the academic ranks, he also had the opportunity to travel. While it was in Germany that he honed the skills as a scientific administrator and diplomat that would come to define his later career, it was on his travels that he did some of his most interesting research. "This time characterises to some extent my whole scientific life," he recalled, "I moved up in energy every 10 years or so by a large factor, and this meant that at each step I had to re-establish myself in a new scientific community."

Herwig had begun his academic career working in optics, in part because nuclear physics was not allowed in Germany in the immediate aftermath of the war, and partly because, as a student at the time, his research options were tightly constrained by the interests of his supervisor. Particle physics did not yet exist as a discipline and nuclear physics was still in its infancy, but his groundings in optics would set him on the path to a research career devoted to the exploration of the fundamental constituents of the universe, and the laws governing their behaviour. As Herwig explained, "in physics, we explore nature, looking deeper and deeper into the microcosm. The smaller the things you want to look at, the shorter the wavelength you have to use, and shorter wavelengths mean higher energy."

The Visible Spectrum and Beyond

The field of optics in which Herwig cut his research teeth essentially relies on photons of low energy, visible light. The objects that can be studied with visible light cannot be smaller than the shortest wavelength that our eyes can perceive, typically around 380 nm, which corresponds to violet light. That's roughly the size of a large virus, so it's pretty small, but by comparison with the subatomic, and subnuclear worlds to which Herwig would later travel, it's huge. An atom is about 1000 times smaller than

© The Author(s) 2024

H. Schopper and J. Gillies, *Herwig Schopper*, Springer Biographies,
https://doi.org/10.1007/978-3-031-51042-7_5

a typical virus and the nucleus of an atom is 10,000 times smaller still. The protons and neutrons, building blocks of atomic nuclei, are a further ten times smaller than nuclei, and their constituents, quarks, are so small as to be considered point-like, with no discernible size at all.

We can't see objects such as viruses with the naked eye, but optical magnifying techniques allow us to do so. To see things that are smaller, we need to go beyond the violet to ultraviolet, followed by x-rays and gamma rays, while at the other end of the spectrum, beyond what we can see, lies the infrared. "Max Planck told us that the smaller the wavelength of a photon of light, the greater its energy," explained Herwig. "In physics one of the units used to measure energy is the electron volt, eV, literally the energy required to move one electron through a potential difference of one volt. The energy carried by even the most energetic of visible photons is only about 3 eV, which is also the kind of energy a torch battery provides to accelerate an electron to this energy."

The current experiments at CERN's Large Hardon Collider (LHC) probe matter with an energy measured in TeV: tera, or trillion, electron volts. This allows structures deep inside the nuclei of atoms to be resolved, but when Herwig started out, such energies were unimaginable anywhere except in the cosmic rays pervading the universe, and constantly bombarding the Earth. The LHC effectively behaves like a microscope that can explore objects about 10^{15}, that's a one followed by 15 zeros, times smaller than anything visible with an optical instrument.

It is difficult to pinpoint the precise origin of modern physics, but a good candidate is 1834, when Michael Faraday, working at the Royal Institution in London, demonstrated that the atoms making up every chemical element carry with them an electric charge that is an integral multiple of that carried by the hydrogen atom, indicating some commonality between the elements. Then, in 1879, Russian physicist Dmitry Mendeleev published the periodic table of the elements. In it, he classified all the known elements in order of their mass, and organised according to their properties. This proved to be a milestone in humankind's understanding of the microcosm, allowing scientists to make predictions based on the observations of patterns that emerged from the natural order. Gaps in the table pointed to yet-to-be-discovered elements, while the underlying patterns pointed to some deeper substructure within the atoms, which remained to be discovered. One could say that Mendeleev's insight marked the transition from chemistry to modern physics, and it was not long before evidence to support the substructure theory was found.

In 1895 Wilhelm Conrad Röntgen, working at the University of Würzburg in Germany, discovered *eine neue Art von Strahlen*—a new kind of rays. These rays bear his name in the German language, and are known as x-rays in English. Just like visible light, they are a form of electromagnetic radiation, but with an extremely short wavelength. A little over a decade later, in 1906, the British physicist Joseph John Thomson conducted an experiment that provided evidence for the first subatomic particle, the electron. Thomson had discovered the particle responsible for the observations of Faraday and Röntgen, and at the same time, established the field of particle physics long before it had a name. Röntgen was awarded the inaugural Nobel Prize for Physics in 1901, while Thomson received the prize in 1906. Further discoveries were soon to follow, notably that of Ernest Rutherford and his graduate students Hans Geiger and Ernest Marsden in 1911.

In the final year of the nineteenth century, Rutherford, working in Manchester, discovered what he called alpha and beta rays in the radioactive disintegration of uranium nuclei. In 1911, he designed an experiment to put the alpha rays to good use. At the time, the atom was famously modelled as being rather like a spherical plum pudding, with a mass of positive electric charge dotted with negative charge to give the atom overall neutrality. Thomson's experiment had demonstrated that the negative charge was composed of electrons, but Rutherford wanted to find out about the rest of the atom, and in particular whether it really was built like a plum pudding. "It became clear at the beginning of the twentieth century," explained Herwig, "that one could not see the details of atoms with normal light, so what do you do? Let's say you have a fog, and you want to explore what's happening in the fog. You can't see into the fog, so you shoot particles or bullets, whatever you want, across the fog. If the bullets go through undisturbed, you conclude the fog's empty, but if you find a few are deflected, by observing these deflections you can conclude what the structure of the objects inside the fog is." In Rutherford's experiment, the fog was a foil of gold, and the bullets were alpha particles. Geiger and Marsden saw that most of the alphas went straight through, but a small number were deflected, sometimes through very large angles. By analysing these deflections, they were able to conclude that the atoms of gold were mostly empty, but that at their core was something much smaller than the atom—the nucleus. "The nucleus is surrounded by shells filled by electrons that circulate around the nucleus," said Herwig, "but most of the atom is empty, most of its mass is concentrated in the tiny nucleus at the centre."

It was not long before the constituents of the nucleus had been identified. Positively charged protons, which balance the charge of the electrons in atoms, were discovered by Rutherford in experiments carried out between 1917 and 1919, and the electrically-neutral neutron by James Chadwick in 1932. It had been a journey of almost a century since Faraday, but these discoveries meant that by the time Herwig was discovering the field of physics at the University of Hamburg in the late 1940s, the foundations of nuclear and particle physics were firmly in place, although tantalisingly out of reach to researchers in post-war Germany, where investigations of the nucleus were forbidden.

A Sojourn in Stockholm with Lise Meitner

"In order to learn nuclear physics, one had to go somewhere else," explained Herwig. "Fortunately, Fleischmann, my Professor at Hamburg, had an opportunity to propose me for a one-year Fellowship in Stockholm, which I accepted with pleasure." The year was 1951, not long after Herwig's marriage to Ingeborg, and even though moving would mean leaving his new bride behind in Hamburg, the opportunity to work with one of the pioneers of experimental nuclear physics, Lise Meitner, was one that Herwig could not refuse.

Austrian by birth, Lise Meitner is one of the least well-known greats of physics from the first half of the twentieth century. Known mostly for the fact that she did not share the 1944 Nobel Prize with Otto Hahn, her contributions to physics are as remarkable as the story of her life, which Herwig was to learn first-hand in Stockholm. "She was able to leave Berlin just in time," recalled Herwig.

Meitner had been invited to Sweden in June 1938 by Manne Siegbahn, a Nobel Prize winner and director of the physics department of the Nobel Institute of the Royal Swedish Academy of Sciences in Stockholm. It was literally a lifeline, although she had an agonising wait to find out whether Sweden would allow her to enter: with the Anschluss of Austria in March of that year, her passport was no longer valid. She escaped Germany by the narrowest of margins. While her colleagues at the Kaiser Wilhelm Institute in Berlin maintained the pretence that she had travelled to Vienna to visit relatives, she was smuggled out of Germany in early July into the Netherlands via a little-used railway crossing, and there she waited to find out whether she would be allowed to enter Sweden and take up Siegbahn's offer. The good news arrived before the month was out, and she travelled on via Denmark, where she stayed with Niels and Margarethe Bohr, to Sweden, arriving at the beginning of August.

Already in her 60s, Meitner was close to retirement when she arrived in Sweden. "I think that Siegbahn thought that she'd settle down, have a quiet life, follow what's going on in physics but not be an active researcher anymore," said Herwig. "She was not expected to carry out any experimental activity in his institute, so after a short time, she decided to accept an invitation from the *Kungliga Tekniska Högskolan* (KTH), the Royal Institute of Technology, under Gudmund Borelius, where she could set up a modest experimental laboratory and continue her experimental work. I joined her there."

Life in Stockholm was a revelation to Herwig, whose entire adult life had been lived under the influence of war or its aftermath. "Germany was still in ruins when I went to Stockholm," he recalled, "and we just had enough to eat. The contrast was incredible. Sweden had not been involved in the war, and it was a paradise. There was enough food, no shortages at all. I enjoyed going to the opera and to concerts, and one thing that was incredible for me to see was that the street-sellers of *varmkorv*, hot sausages, were dressed in furs against the cold. In Germany, they were still in rags. Incredible." One thing, however, was in short supply in Sweden: alcohol was rationed. This too contributed to the good life that Herwig was living. "The fixed ration for foreigners was relatively high, so I could buy spirits, and whenever I was invited to a party, I took a bottle along so I was always a welcome guest. I enjoyed life in Sweden."

Life was not all fun and games, however, there was also serious work to be done, and Herwig hit it off well with his new supervisor. "With my Austrian roots, we had something in common," said Herwig. "Apart from the physics, we immediately had a very close exchange of our personal histories, and she was happy to have someone to talk to about her early years. She told me her life story, and the difficulties she'd faced first as a woman and then as a Jew." Before the First World War, it was very unusual for a woman to study physics, but Lise Meitner was fortunate in her teachers, and in 1906, she became one of the first women to earn a doctorate from the University of

Vienna. "After her doctorate, she went to the Kaiser Wilhelm Institute, as it was called at the time, now it's the Max Planck Institute, in Berlin. She went to the institute for radio chemistry where the director was Otto Hahn. They immediately became friends, but he could not do what he wanted and hire her in the normal way: women were formerly forbidden at that time to enter the institute, so she was accepted at the institute more-or-less secretly. She told me that in the beginning she was not allowed to enter through the main door and had to use a side door. Nor was she allowed to use the main laboratory in the institute, so Hahn had established a little laboratory in the basement in what had been a carpenter's workshop, and it was there that she worked with Hahn."

Lise Meitner had a good ear, which was put to good use in her relationship with Hahn. Both music lovers, they would sing together the famous lieder of Johannes Brahms as they worked, and Meitner was quick to criticise when Hahn was off-pitch. The two formed a strong working bond, he a nuclear chemist, she a physicist, at the time when the boundary between the fields was blurred and the talents of both were needed to interpret the results of experiments. Their partnership was cut short by the rise of Nazism, which forced Meitner's escape. "She told me that she got all the help she needed from Hahn to get safely out of Germany. He accompanied her to the railway station, and gave her a diamond ring as a last resort in case she needed money."

Hahn and Meitner's greatest triumph came when she was in exile in Sweden. Back in Berlin, Hahn had been conducting experiments bombarding uranium nuclei, the heaviest to have been observed at the time, with neutrons in a bid to create heavier, so-called transuranic, elements. What they found was the opposite of what they expected, the lighter element barium appeared in the reaction products. "They found lighter elements and couldn't explain it", said Herwig. "So Hahn wrote a letter to Lise Meitner explaining his experimental results and asking for her help interpreting them. She got that letter just before the Christmas of 1938, which she was spending with her nephew, Otto Frisch, who was also a physicist. When the letter arrived, he was putting on his skis to go cross-country skiing, so with the fluttering letter in hand, she ran after him and called him back. He was reluctant, but she persuaded him. He took off his skis and when they sat down to discuss Hahn's results, they had the idea that it must be nuclear fission, which nobody had expected." In 1933, Frisch, also Austrian by birth but working in Hamburg at the time, had left Germany for London. When he and his aunt interpreted Hahn's results, he immediately informed his colleagues there of the news. "Because of the political situation in Germany, Meitner and Frisch published separately from Hahn, with the latter's paper describing the experiment, while Meitner and Frisch explained the physics behind it."

Coming on the eve of the Second World War, the result caused a sensation. It showed that the atomic nucleus could be split, with all that implied. Frisch went on to work out, with Rudolf Peierls, the process for generating a nuclear explosion, and as a freshly minted British citizen, left to join the Manhattan Project in 1943. Hahn was awarded the Nobel Prize for Chemistry in 1944, as the war was drawing to a close. There was no award ceremony that year, and he was formally informed of the news while detained at His Majesty's pleasure at Farm Hall, where many prominent German scientists were interrogated after the war.

"There were some rumours that Meitner and Frisch missed out on the Nobel Prize due to negative interventions from Hahn," said Herwig. "She told me that's just not true—Hahn had always recognised that as a nuclear chemist, he needed the expertise of physicists to interpret his results. I checked that later, many years later, when I met Hahn. I asked him if it was true. He said, 'Of course, Lise was the only one who understood what we were doing. The Nobel files for the 1944 prizes are now open, and anyone can study the deliberations of the chemistry prize committee.' The jury is still out as to why Meitner and Frisch were overlooked, but among the most likely explanations is that the chemists simply did not understand the importance of the physicists' contribution."

When Herwig arrived in Stockholm, all this was still recent history, but however she may have felt about it, the omission did not distract Meitner from her research, and Herwig was about to take his first big step up in energy. "Because the possibilities were relatively limited in Stockholm, she went back to some previous work that she had been doing," recalled Herwig. "She had been quite essential in the discovery of the neutrino, much earlier than nuclear fission, when she was investigating nuclear beta decay."

In 1930, Meitner had been in an unrivalled position to witness the birth of neutrino physics. She, along with Hans Geiger, was the recipient of a famous letter from Wolfgang Pauli, which opened: "*Liebe Radioaktive Damen und Herren*" (Dear radioactive Ladies and Gentlemen). Meitner and Geiger were involved with a physics gathering in Tübingen, and Pauli wanted to sound out the participants on a new idea he'd had.

One big dilemma in physics at the time was related to beta decay. We know now that beta decay happens when one of the neutrons in the nucleus becomes a proton. An electron is emitted in the process, ensuring that charge is conserved. In 1930, however, the neutron's discovery was still two years into the future, and the dilemma was that the electrons emerged with a range of energies in a way that seemingly defied the sacrosanct principle of conservation of energy. "Meitner had already contributed much to the study of beta decay, demonstrating that the spectrum of emitted electrons was continuous," explained Herwig, "and she had conducted a rather difficult experiment that showed that some of the energy was missing."

Pauli's solution to this dilemma was to propose that the decay involved an extra, neutral particle that escaped undetected, thereby accounting for the missing energy. He called this particle the neutron. With the discovery of what we now know as the neutron by James Chadwick in 1932, however, Pauli's neutral particle was in need of a new name. Thanks to Enrico Fermi, it is today called the neutrino, or 'little neutral one.'

Meitner's indirect evidence for the neutrino had inspired Pauli's missive, but direct evidence would have to wait until the 1950s. In a sign of how parsimonious Nobel committees can sometimes be, the physics prize for its discovery was only awarded to one of the experimenters, Frederick Reines, in 1995, by which time his co-researcher Clyde Cowan had passed away.

When Herwig arrived in Stockholm, the elusive neutrino was still evading detection, and Meitner was still investigating the finer details of beta-decay. "The task she gave me was to measure the beta spectrum by putting different absorbers in and measuring how many electrons with a certain energy passed through the absorbers," said Herwig. "The first thing she taught me was to build a Geiger–Müller counter and then to set up an experiment with a radioactive source, an absorber and then the Geiger–Müller counter to measure the counting rate with different absorber thicknesses. It was an incredibly simple experiment compared to the hugely complicated magnetic spectrometer installations of today!"

Meitner was an exacting teacher. "She was critical in setting up the right geometry for the experiments," Herwig recalled. "She told me she was famed for her good eyesight, and had been the scourge of dressmakers back in Vienna, where she could see at a glance whether the hem was straight." And so it was with Herwig's experiments. "She'd say: 'Here it's not my hem but your experiment, that's not aligned properly!' So I learned how to do an experiment properly and we published a paper." It was not a great discovery, but it was a solid piece of nuclear physics, and it represented Herwig's first big step up in energy from the few electronvolts of optics to the tens or hundreds of thousand electronvolts, keV, of beta decay and nuclear physics.

Herwig's stay in Stockholm was just a year, but he remembers it fondly. "I shared an office with a Swedish scientist, and I think we were the last scientists who worked

Fig. 5.1 Lise Meitner in her laboratory in Stockholm around 1950. During Herwig's year in Stockholm, Meitner introduced him to experimental nuclear physics (©Department of Physics, KTH Royal Institute of Technology, Stockholm, All rights reserved)

with Lise Meitner, because after I left she very soon retired from the practical work," he said. "So I am probably the last living physicist who still really joined her in her work."

Without his wife, and despite his popularity at parties, Herwig spent most of his time in Stockholm working. "I lived relatively modestly and was able to save some money, so at the end of the academic year, I had saved enough to invite my wife to come and join me in Sweden. I could pay for her trip and support her stay for a few weeks."

Ingeborg, who had stayed in Hamburg, still working for the British Military Government there, joined Herwig for his last month in Sweden in the summer of 1952. "I had been there through the whole winter, which was somewhat depressing – the long nights, no sun. I was not accustomed to that," remembered Herwig. But with the summer came Ingeborg, and on 21 June, midsummer night, a big festival across the whole country. "I was invited with my wife to Lund University, and I was very much impressed that all students were in evening dress. In Germany no student would have appeared in evening dress in front of the rectorate. The rector gave a speech from the balcony and at the end of the speech they all threw their white caps, which are the identification of students in Sweden, into the air to mark the beginning of summer. It was a very nice and very impressive festival, and a fitting way to end my year in Sweden, where I got an introduction to nuclear physics, into another layer of the microcosm, but also an introduction to another culture. This was a very important stage in my life."

On to Cambridge

It was not long after Herwig got back to Hamburg that he followed Rudolph Fleischmann to Erlangen, where he continued the work he'd begun with Lise Meitner in Stockholm. Before long, Fleischmann offered Herwig the chance of another sabbatical, this time in England. "Fleischmann was interested in nuclear physics, and he said: 'if you go to England, to Cambridge, you can go to the famous institute where nuclear physics started with Rutherford at the beginning of the twentieth century and learn how to do nuclear physics with an accelerator.' With Lise Meitner I had learned about radioactive decays, beta decay, but Fleischmann knew that the future lay with accelerators—at Heidelberg, he had worked with Walther Bothe, the Nobel Prize winner who had invented the coincidence method of detecting two particles emitted at the same time."

In 1947, Lise Meitner's nephew, Otto Frisch, had taken up a position as director of the nuclear physics department of Cambridge University's Cavendish Laboratory. The department had acquired a modern version of a Van de Graaff accelerator operating at about 3 mega electronvolts (MeV), another step up in energy for Herwig. This time, conditions were sufficiently comfortable for the Schopper family that Ingeborg and Doris were able to join Herwig soon after he had arrived in Summer 1956 (see this chapter, In his own words: Learning about the English way).

"The scientist who was running the Van de Graaff was Denys Wilkinson. He was a young shooting star, quite well known already in Great Britain, and he was kind", recalled Herwig. "I started an experiment at the Van de Graaff investigating the splitting of the deuteron, a nucleus consisting simply of a proton and a neutron, using energetic photons. I worked there for a few months, and in the end a publication came out, but what happened next was completely unexpected."

Herwig had gone to Cambridge to learn about accelerators, and that's how things started, but soon he became friends with Otto Frisch through the shared love of music that came with their Austrian heritage. Herwig was a regular visitor to the Frisch home, where he'd spend evenings listening to Otto play the piano. One day, Otto told him of an upcoming colloquium at the Harwell campus in Oxfordshire, and suggested that Herwig should attend. It was to be a life-changing experience. "The speaker was Abdus Salam, a Pakistani theorist who was already very well known," recounted Herwig. "He was a professor at the University of London, and he received a Nobel Prize much later."

Salam shared the 1979 Nobel Prize for Physics with the Americans Sheldon Glashow and Steven Weinberg for the unification of two of nature's fundamental forces, electromagnetism and the weak nuclear force into a single electroweak theory. Back in the 1950s, however, his attention was firmly focused on the weak interaction itself, which is responsible for beta decay and plays a crucial role in energy production in the sun. Under different circumstances, Salam might have received science's most coveted prize much sooner, as Herwig was to learn that day.

The chairperson of the colloquium was Wolfgang Pauli, who came to Harwell from his home in Switzerland just to chair this colloquium. Pauli was famed not only for his intellect, but also for his self-confidence: Pauli never apologised to anyone for being wrong, simply because he never was. At Harwell that day, however, a sense of humility was on display from this giant of physics.

"I learned something new that day," said Herwig. "There was a concept of physics that I learned at this colloquium for the first time. Among the fundamentals of physics are some principles that were considered to be necessary for rational thought, '*denknotwendig*' in German." Among these is the one that Herwig learned from Abdus Salam: the concept of symmetry, which goes to the heart of modern physics, and has some very far-reaching consequences. "The concept of symmetry means that laws of nature do not change under certain operations," explained Herwig.

The Principle of Symmetry Invariance

"The most fundamental symmetries are based on the idea that the results obtained by an experiment should not depend on the particular place where the experiment is carried out," explained Herwig. "They should also be independent of the direction in which the experimental apparatus is oriented, and they should not be influenced by the moment at which the experimenter starts their watch. The full importance of these fundamental symmetries was first pointed out by Emmy Noether, who published a paper in 1918 showing that from each of these symmetries a conservation law of physics follows, namely the conservation of linear momentum, angular momentum and energy from the three I've just mentioned."

Similar consequences follow from mirror reflections. If we observe nature through a mirror it looks exactly the same except that a left-handed glove becomes a right-handed glove and vice versa. "The laws of classical physics do not contain a definition of handedness," said Herwig, "so there is no way to explain to an alien on a distant planet what a right-handed glove is. The only way would be to send one."

Symmetries in physics seem very abstract, but they have very tangible consequences. For example, we know that as well as ordinary matter of the kind that we are made of, there also exists antimatter in the universe. Rather like the concept of handedness, what we call matter and what we call antimatter is arbitrary. "The only way to find out whether two pieces are of the same kind or not is to bring them together, since matter and antimatter annihilate on meeting," explained Herwig. Indeed, observations indicate that most of the matter and antimatter that would have emerged from the Big Bang have annihilated, leaving just a small fraction of the matter behind: imperfect symmetries turn out to be important in physics as well.

Another symmetry is time reversal. This might seem surprising as we observe time always flowing in one direction—we get older, never younger—but all the laws of classical physics remain valid if we reverse the flux of time. "If one could make a film of the motion of the planets around the sun and show it to an astronomer," said Herwig, "they would certainly be able to deduce Kepler's laws, but they could not tell you whether the film was being shown forwards or backwards in time." To explain why we get old, classical physics was obliged to introduce another concept, entropy, defined in such a way that it can only increase when moving to the future.

An Early Experiment on Mirror Reflection Invariance

The symmetry Herwig learned about at Harwell is mirror symmetry, along with its consequences for quantum theory. In physics, mirror symmetry goes by the name of parity. "We think that when you do experiments, nature can't tell the difference between the observation and its mirror image—the laws of nature are the same for both," said Herwig. Principles of symmetry have intrigued thinkers since time immemorial, you only have to look at nature to find them—in the human form, in the petals of flowers, or in rock crystals. For Herwig, the philosopher Immanuel Kant is a reference point. "Kant said that a necessary principle to investigate nature is to assume that there is a mirror invariance that nature cannot decide between, for example a right-handed glove is a mirror image of a left-handed glove", he explained. "The same is true for nuts and bolts, there are right-handed bolts and left-handed bolts. A right-handed bolt will never fit into a left-handed nut, but this does not imply that left-handed nuts cannot exist, it just says the laws of nature must be invariant as far as mirror transitions are concerned."

Surprising as it may seem, this principle translates directly to the world of quantum mechanics and the fundamental particles, which have a property called helicity, which can be right or left-handed. "In quantum mechanics, the equivalent principle is called helicity invariance," Herwig continued, "and there is a law in quantum mechanics that if you have an invariance, it comes with a quantum number that does not change in reactions. This quantum number is called parity, and I taught myself all there was to know about it after this colloquium when I returned to Cambridge."

Before 1956, physicists believed strongly that all laws of nature must be mirror invariant: that everything that could happen in physics would be identical in a mirror image. But Abdus Salam had come up with a two-component theory of the neutrino, which said that in beta decay, the emitted neutrinos must have a particular helicity, or handedness, whether in our universe or a mirror one, although his theory could not predict which helicity is preferred by nature in our universe.

Salam's theory violated parity conservation, and at the beginning of the collo-quium, Pauli, who never, never, admitted that something he had said was wrong, excused himself for having persuaded Abdus Salam not to publish his two-component theory, because some rumours were going around that an experiment had shown that parity conservation might be violated. "That's what really caught my attention that day," said Herwig. "Salam would probably have had the Nobel Prize if he'd published. As a consequence of that colloquium, my research took a new turn, back in the direction of beta decay." What Herwig did not know when he set off for Harwell is that across the Atlantic there had been feverish activity, both theoretical and experimental, in elucidating the weak interaction.

As 1956 drew to a close, weak interaction physics was undergoing something of a revolution. A paper published in the summer by two Chinese-American theorists, Tsung-Dao Lee and Chen-Ning Yang, interpreted results from cosmic rays and from the Brookhaven National Laboratory in the US, as perhaps being an indication that parity was not conserved in weak interactions. Their paper did nothing less than challenge one of the basic tenets of physics, although in the summer, there was no watertight experimental evidence to show that they were right.

Lee's home institute was Columbia University in New York City, where he knew the experimentalist Chien-Shiung Wu, who already enjoyed a powerful reputation. Wu had moved to the US from China in 1936, and worked on the Manhattan Project through the war, not to mention her experimental forays into beta decay, and she was the first woman to become a professor of physics at Columbia. Lee had discussed the paper he was about to publish with Yang with her before it was published, and by the time Herwig was listening to Salam's words, she was already turning her formidable experience to providing the first definitive experimental proof that Lee and Yang were right.

"In their paper, Lee and Yang proposed four experiments to check whether mirror symmetry was true or not in weak interactions," explained Herwig. "They proposed one experiment in beta decay which Mrs Wu immediately started to do, and in the autumn of '56, she had her first results. Her paper had not yet been published, but there were rumours that her results showed that parity invariance was wrong." Those were the rumours that had brought Pauli to Harwell, where he said that if they were true, he owed Salam an apology. They were still just rumours, though, and Herwig remembers Pauli expressing scepticism, drawing on the Kantian ideal that reason must be based on principles of symmetry.

Herwig's interest had nevertheless been aroused: here was an opportunity to make a major contribution to physics. When he got back to Cambridge, he read Lee and Yang's paper, where he learned about the four experiments they proposed. "Two were experiments in nuclear physics, in beta decay, and one was the one that Wu was doing. The second predicted that if a beta decay is followed by a gamma emission, the emitted photons must have a definite helicity which means they would be circularly polarised." Lee and Yang thought that this experiment could not be done, because they were not aware of any experimental method to measure the circular polarisation of the photons. Herwig, however, knew better. "When I read that, I jumped up because, before coming to Cambridge, I had done exactly that at Erlangen, in an experiment trying to measure the helicity of gamma rays."

When Herwig reported back to Otto Frisch and asked whether he could do the experiment at the Cavendish, the answer he got was an enthusiastic yes. Herwig stopped his work on the Van der Graaf, and got back to work on beta decay. "After Christmas, I immediately started to do this experiment, and Frisch gave me all the support I needed." Herwig's work in Erlangen had shown that circular polarisation of gamma rays could be measured by scattering the gamma-ray photons from a magnetised iron cylinder. Although Herwig was given priority at the Cavendish lab's workshops, he decided to build the apparatus himself to save time, winding the copper coil around the cylinder by hand, and feeling grateful for his early training in Hamburg that had taught him all the manual techniques that modern-day physicists don't have to concern themselves with. The other essential ingredient for the experiment was a radioactive beta-emitting source. "Wu had done her experiment with a cobalt-60 source, and the beta decay in this nucleus is followed by a gamma emission," said Herwig, "so I asked Frisch to get me such a source."

As a fellow of the Royal Society, Frisch was very influential in the UK, and he was able to procure the source within a week. "With a cobalt-60 source, the same nucleus that Wu used, I started to do the measurements within a month, by the end of January, a record time for setting up a nuclear physics experiment," said Herwig with a smile. "I had my first results in February, and they were published immediately." The rapid publication time also owed much to Otto Frisch's status in the UK nuclear physics community. "He was editor of *The Philosophical Magazine*, and he made sure that my paper was published within a week." Herwig's paper, 'Circular polarization of γ-rays: Further proof for parity failure in β decay', was received on 14 March 1957, and published the same day. C. S. Wu's paper had appeared in the American journal *Physical Review* just one month earlier, on 15 February.

"My paper appeared in *The Philosophical Magazine* only a few weeks later than the paper of Wu, but of course, American physicists didn't read *The Philosophical Magazine*, so it took several months until people became aware of my paper," said Herwig, in a clear illustration that the centre of gravity of fundamental physics had moved yet further to the west. Nevertheless, the paper did eventually get him noticed. "Within half a year, it really put me on the landscape of international nuclear physics,"

Circular Polarization of γ-rays :
Further Proof for Parity Failure in β Decay

By H. Schopper
Cavendish Laboratory, Cambridge†

[Received March 14, 1957]

Lee and Yang (1956) suggested several experiments for testing the conservation of parity in weak interactions. Two of these have been performed (Wu *et al.* 1957, Garwin *et al.* 1957‡) and have shown that parity is not conserved. Results of a third experiment (thought impracticable by Lee and Yang) are reported here. They confirm the expectation that the γ-rays emitted after β-decay at an angle θ relative to the β-particle should show circular polarization proportional to cos θ.

† Permanent address : Physikalisches Institut, Erlangen, Germany. This work was performed during a visit to the Cavendish Laboratory.
‡ The author would like to thank Dr. D. H. Wilkinson for providing preprints of these papers and for other valuable information.

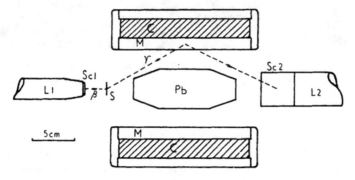

Fig. 5.2 Circular Polarization of γ-rays: Further Proof for Parity failure in β Decay: Herwig's paper demonstrating Lee and Yang's theory of parity violation. Published in *Philosophical Magazine* in May 1957 [1], the paper is remarkable in many ways, not least that it is a single-author paper, which would be unheard of in modern-day particle physics. It remains the only paper showing that neutrinos and antineutrinos have opposite helicities using a single experimental set-up as shown (H. Schopper (1957) A Journal of Theoretical Experimental and Applied Physics, 2:17, 710–713 ©Taylor & Francis license, All rights reserved)

he recalled, "That's how I became known in the international area of nuclear physics, thanks to Frisch and thanks to this experiment."

Herwig went on to repeat the experiment with a source that emitted positrons, the antimatter equivalent of electrons, followed by a gamma, and the result still stood. "I was able to show that the circular polarisation of the gamma rays following electron and positron beta decay have opposite helicities," he said. "Still today, I think my experiment is the only one that has been able to show this with the same experimental equipment, just by changing a radioactive source. I'm very proud of that! And remarkably, my paper had only one author."

Fig. 5.3 Herwig discussing helicity with Chien-Shiung Wu. This photograph was probably taken at a conference in Geneva in 1958 (Herwig Schopper's personal collection. ©Herwig Schopper, All rights reserved)

The revelation that parity was not conserved once again showed the vagaries of the Nobel Prize. The 1957 Nobel Prize for Physics was awarded to Lee and Yang. Chien-Shiung Wu had to wait until 1978, when she was the inaugural recipient of the Wolf Prize, for recognition of her experiment. As time progressed, she was not silent on this omission, pondering, for example, in a 1964 MIT Symposium: "whether the tiny atoms and nuclei, or the mathematical symbols, or the DNA molecules have any preference for either masculine or feminine treatment." But Herwig thinks there might have been another reason for this omission. "In order to carry out the Wu experiment the nuclear spins of the radioactive nuclei had to be oriented in one direction, which required extremely low temperatures not available at Columbia University," he explained. "Hence she had to do the experiment with a cryogenic group in Washington and the paper was signed by more than three authors. The Nobel can be given only to three scientists and not to a group—a too restrictive rule for modern sciences."

As time progressed and Herwig's renown grew, he was invited to give talks at the American Physical Society and many other conferences and colloquia. "I got to know Lee and Yang—Lee became a friend," he recalled. "I became good friends with Wu, and we met very often at conferences. In fact, in 1957, something very special happened to me, I was invited to an international conference in Israel together with

Lee and Wu to talk about our results, and I think I was probably the first German to be invited to an Israeli conference after the war, so that was a big moment."

Looking back on his time in Cambridge, Herwig is led to some philosophical musing on the nature of knowledge, and our interpretation of results. "It turned out these experiments not only show that parity, the mirror symmetry, is violated, but also that the symmetry between matter and antimatter is violated," he began. "I could show, at least with the same accuracy, and I think with even better accuracy than the Wu experiment, that this symmetry is also violated to the maximum possible degree theoretically allowed. When it became known that parity and matter–antimatter symmetry are both violated, theorists took that for granted automatically.

Fig. 5.4 In 1988 Herwig was presented with the Weizmann Institute's Gold Medal by former French Minister and President of the European Parliament, Simone Veil, Honorary President of the Pasteur-Weizmann Council, in recognition of his work to strengthen relations between CERN and Israel. On the right is J.-J. Brunschwig, Chair of the Institute's Suisse-Romande delegation (©CERN, All rights reserved)

But if you consider the now famous standard model of particle physics, this parity and charge symmetry violation is put in by hand. There's no fundamental theoretical reason for the sign of helicity of the neutrino, it's only the experimental measurements that justify it. Today, nobody talks anymore about parity violation, it's taken as given, but people are still talking about matter–antimatter asymmetry, although both were put into the standard model by hand at the same time. But then people were surprised that matter–antimatter symmetry seems to be violated in the cosmos, as well as at the microscopic scale. In the cosmos, it seems that all galaxies consist of matter and not of antimatter. So people are getting excited about the matter–antimatter violation, but they are not excited anymore about the mirror equivalence violation. To me, this shows a strange behaviour of scientists concerning 'revolutions' in physics. Sometimes in physics, in theoretical physics, some fundamental concepts are taken as given, whereas others are not, although there are no clear theoretical arguments in favour of one and against the other. Many discussions between physicists and philosophers could be saved if one would accept such arbitrariness and agree that the experiments show what is realised in nature: logical thinking alone is not always enough. But let me emphasise that I still believe that a close co-operation between theory and experiment is necessary for any progress in science. The theories sometimes present a question and leave the experiments to decide how to answer it, but sometimes by pure chance, a new phenomenon is discovered without having been predicted. Both approaches are necessary to gain basic new knowledge."

His year at Cambridge done, in autumn 1957, Herwig packed his bags for Erlangen, and took his new research interest with him.

A Year at Cornell

Back in Erlangen, Herwig introduced his students to beta decay, but the following year, 1958, he was off to Mainz to take up a new post. The university there was relatively young, and its physics department was mainly occupied with teaching. It was headed by Rudolf Kollath, who Herwig knew from Hamburg. In order to introduce more research, there were plans to create an institute for experimental nuclear physics. Herwig's job, as well as becoming a professor, was to set up and direct this new institute. Although moving towards science administration, Herwig managed to keep his hand in with research. "We continued the nuclear beta decay experiments with some collaborators who had come with me," he recalled. "We produced some interesting and detailed results that helped understand the weak interaction" (see Chap. 4, Moving to Mainz and the foundation of MAMI).

Fig. 5.5 Mainz's young
Professor Schopper made an
appearance in the *Fuldaer
Zeitung* newspaper in
September 1958. He'd been
invited to present the results
of the experiment he
conducted at Cambridge on
parity violation at the second
Atoms for Peace conference
in Geneva in 1958 (©Fuldaer
Zeitung, All rights reserved)

It wasn't long, however, before an opportunity arose to plunge right back into new research. Fleischmann's successor in Hamburg, Willibald Jentschke, needed someone to learn the art of doing physics at an electron machine in order to establish an outside user group for the new *Deutsches Elektronen-Synchrotron*, DESY, laboratory that Jentschke was setting up in Hamburg. He asked Herwig to go to the States for a year, and Herwig was keen to accept. He deferred the post he'd been offered at Karlsruhe, and in August 1960, the Schopper family, now four in number with the arrival of Andreas in October 1959, was once again on the move. Jentschke had arranged for Herwig to receive a Fellowship to work with Robert Rathbun Wilson, founder and director of Cornell University's Laboratory for Nuclear Studies, which was home to a 1.4 billion electronvolt, or GeV, electron synchrotron at Ithaca, in upstate New York. At the time, it was the largest electron machine in the world.

"So I quit my job in Mainz, accepted the chair at Karlsruhe but started with a year's leave of absence without pay," explained Herwig. "My income would come from the Fellowship, so that's what I did, and Bob Wilson happily accepted me. He had arranged the fellowship from the National Science Foundation. It was a lousy fellowship, I think it was $500 a month for the academic year, but I thought we could live there modestly on that." The next hurdle was to obtain an immigration visa that would allow him to work in America. "It was very difficult for a German to get a visa for the USA," said Herwig. Immigration visas to the US were rationed, and there was a quota for each European country. At the time he applied, the German quota was exhausted, but there were practically no Czechoslovaks who could immigrate to the United States. "I was lucky in this sense because I was born in Czechoslovakia, and the rules in the United States were, and still are, that to get a visa, your birthplace is decisive and not your actual citizenship." So Herwig was eligible for an immigration visa on the Czechoslovak quota, but there were still more hoops to be jumped through. "I had to fill in a form and appear before a diplomatic representative at the American embassy, and I had to solemnly swear on all kinds of things."

Herwig duly presented himself at the nearest US diplomatic mission in Frankfurt and was presented with a long form to complete. "There were about two dozen questions that I had to answer, and after each question I had to swear. Would I respect the Constitution of the United States was one of the first questions. Then I was asked to swear that I would not murder the President, or organise prostitution in the United States, and so on. There were many, many questions, and I had to swear individually on all of them."

Herwig survived the ordeal, and was awarded a visa, but it was a close thing. "I couldn't help thinking about another physicist, Friedrich Houtermans, who'd told me about his experience of immigration to the United States. When he went to the American Embassy in Switzerland, they asked him the same questions, and after about twelve, he said, "Well, if you have more stupid questions like that, then I renounce my request for an immigration visa." And he did. He didn't get his visa. So when I was going through all these questions, swearing, I was thinking of Houtermans, and I was afraid that I would start to laugh. If I did, they might misinterpret my behaviour and believe I was not respecting the laws of the United States. But I controlled myself and got my visa."

As air transport was still a luxury in the 1960s, the Schoppers boarded a ship in Hamburg, and were in for a choppy crossing. Doris, now five years old enjoyed the adventure, but baby Andreas did not appreciate the trip. Herwig spent much of the time in the ship's pool, employing the kind of thinking that only a physicist could. "It was the first time that I crossed the Atlantic. I got seasick, but when I went into the swimming pool it immediately disappeared since the water did not follow the slow motions of the ship." They arrived in New York after a week at sea, and were greeted, like so many before them, by the sight of lady liberty, lifting her light to guide the tired, the poor, and the huddled masses of the old world into the new. From New York, they travelled to Ithaca by plane, and were very soon installed in a house they'd rented at a very good price from a professor who was away on sabbatical. "It was a beautiful house with a little garden, indeed the first house we could have just for us. This was a great experience for my family."

By the time the Schoppers arrived in Ithaca, Herwig's research interest had moved on to the question of why different forces exist in nature, and what the properties of the fundamental building blocks of matter might be. At the time these were the protons and neutrons of atomic nuclei, since their constituents, the quarks, were not yet known. This required another jump in energy, from the MeV of nuclear physics to the GeV available at the Cornell electron synchrotron. It was a significant jump: 1 GeV is a thousand MeV.

By the 1960s, particle accelerators were firmly established as the research tools of choice for exploring the subatomic, and subnuclear worlds of particles and the forces that govern their interactions. Far from being a purely esoteric field of science, such studies of the quantum world can offer understanding of how the universe works at larger scales, and have gone on to deliver technologies that underpin much of modern society.

Back then, however, the field of particle physics was in its infancy, and Herwig was asking the question of whether the protons and neutrons of atomic nuclei had substructure, just as Rutherford, Geiger and Marsden had shown to be the case for atoms. "Since it had turned out that an atomic nucleus is not a single body but has a complicated structure, why couldn't that happen for the proton and the neutron? Such investigations need smaller wavelengths, higher energies, and that became possible with the development of high-energy accelerators."

Although much of the groundwork for high-energy particle accelerators had been done in Europe, and CERN had started its first big machine, the proton synchrotron, the year before, in 1960, Europe was still playing catch up. The US was the place to be for a physicist in pursuit of the high-energy frontier. Berkeley's cosmotron had been running at its full energy of 3.3 GeV since 1953, and a massive linear accelerator was being planned at Stanford. Two miles in length, the Stanford linear accelerator remains to this day the world's longest linear accelerator. The cosmotron was a proton accelerator, whereas the Stanford machine would be an electron machine, making them complementary in terms of the research they could support.

Herwig, however, was interested in a different kind of machine. The facility that Jentschke was planning for Hamburg was to be a circular electron machine, like the one at Cornell, and Herwig was there to learn how to do physics at such a facility. It was a choice guided by several factors, some political and some to do with the science at hand. On the political side, Jentschke was building a facility that would be complementary to the big proton machine at CERN, according to the principle of the ECFA pyramid. That guided the choice of electrons. "Jentschke's idea was not to imitate the linear electron accelerator at Stanford, but to build a circular accelerator for electrons, even if it was more difficult to get to high energies."

On the physics side, there are many parameters to consider when designing a machine for research. Protons are easier to accelerate to high energy in a circular machine than electrons because they lose much less energy when being forced to follow a circular path. Electrons, on the other hand, produce cleaner collisions. That makes protons the particles of choice when the highest energies possible are required, whereas electrons are better suited for precision studies. "If you want a high-energy electron machine, then a linear accelerator is the facility of choice," explained Herwig, "although they are more expensive than a circular electron synchrotron."

It was a Saturday afternoon when the Schoppers arrived in Ithaca, but Herwig was keen to explore his new workplace, so he left his family at home, and drove to the laboratory. "I'd never seen a large accelerator, so I immediately went there. Every-thing was open, there were no access controls, and I went directly to the accelerator hall. I got in with no problem at all, it was empty, nobody was there—it was Saturday afternoon. I walked around the accelerator, and came across a man sweeping the floor and so I started to ask him questions about life in the institute and some details about the accelerator. He gave me surprisingly informed answers, so after a while I said, "Look, you can't be a janitor here, who are you?" He said, "I'm Bob Wilson." I said, "What? You're Bob Wilson, the director of the institute sweeping the floor on a Saturday afternoon?" "Oh," he said, "it's nothing special, when I was a young researcher, the first time I went to Berkeley, where the director was the famous Luis Alvarez, I found someone sweeping the floor there. I asked him, "Who are you?" "I'm Luis Alvarez," he said, "the director of the institute." This exchange was an eye-opener for Herwig, who was used to the European way of doing things, and an introduction to the more laid-back American approach of the time. "That was first time I met Bob Wilson. He was a fantastic person, and we became good friends. I tried much later to introduce the tradition of floor sweeping by directors at CERN but failed completely."

Bob Wilson founded not just one leading laboratory for particle physics, but two. After his time at Cornell, he went on to establish the National Accelerator Laboratory, today's Fermilab, near Chicago, and he was also a pioneer of hadron beam therapy for treating cancer, pointing out that protons deposit all of their energy at the end of their trajectories, unlike electrons or photons, which deposit energy smoothly along their paths. That makes protons good for pinpointing tumours and avoiding collateral damage to healthy tissue nearby. The GSI laboratory, founded by Christian Schmelzer in Darmstadt, later went on to become a pioneer of this technique, and Wilson's idea is acknowledged in a stained glass window—one of several celebrating the science of the nearby laboratory—in a nearby church dating from the twelfth century in Wixhausen.

Fig. 5.6 Robert Rathbun—Bob—Wilson, director of Cornell University's synchrotron laboratory, introduced Herwig to experimental high-energy physics. Wilson went on to become the founding director of the National Accelerator Laboratory, today known as Fermilab. As well as being an excellent scientist, Wilson also brought his artistic sensibility to the Fermilab site, insisting that the buildings and urban architecture be designed according to a uniform aesthetic style (©Fermilab, All rights reserved)

"Bob Wilson was a great personality, he was not only an excellent physicist, but he was also proud to be a painter and sculptor," said Herwig. "Some of his pieces of art were even exhibited at Harvard University, and even if real artists were indulgent of his artistic efforts he was very proud of them. Bob also cared about the environment when it was not yet a very common concern. He designed the new lab building—the high rise—an extraordinary architectural landmark for the whole region. He also designed the electricity pylons of the laboratory, which have a certain touch of Japanese beauty. In addition, he introduced a breeding herd of bison to the laboratory's prairie land. As the herd thrived, on special occasions, for example large international conferences, delegates would be treated to bison grilled on a large open fire. Bob's wife was also fantastic and very good at art. I was very happy that we became friends, not only from the professional point of view, but also privately."

Finding the director sweeping the floor on a Saturday afternoon was not the only culture shock in store for Herwig. "I was very much impressed by the friendliness and helpfulness of our neighbours," he explained. "When we moved into the house, the neighbours immediately came and asked my wife how they could help us to settle down, so that really impressed us very much." When Herwig started to work at the laboratory and regular night shifts became a way of life, he'd take a nap through the morning and Ingeborg would prepare lunch for when he awoke. After lunch they'd pay a visit to one of the many beautiful state parks that Ithaca had to offer. To Herwig, it seemed idyllic, but to Ingeborg's friends, it was an affront. On one such day, Ingeborg was invited for lunch, but declined because of the arrangement she had with Herwig. "I almost started a revolution when I said I couldn't come to the wives' lunch because I was preparing lunch for my husband," she explained to Herwig. "I met with a complete lack of understanding: they said, 'it's not acceptable that a wife cannot go to a ladies' lunch because she has to prepare lunch for her husband, can't he prepare something for himself?'" It was the beginning of the 1960s, and European and American ways of doing things were worlds apart. "They gave up on me," said Herwig, "but I was still impressed with them, and with the fact that when my daughter went to school, she didn't only have to remember the names and achievements of all the Presidents, but also those of their wives. I discovered that the influence of women in the United States was already much stronger than it was in Europe at the time."

At Cornell, Herwig learned all about the merits of linear versus circular machines, and protons versus electrons as tools for research. "The advantage of electrons compared to heavier particles is that it's easier with them to investigate the structure of a small thing like a proton," he explained. "If you use protons as the probe, you get all kinds of complications because of the complicated inner structure of the proton itself. These complications are very interesting in their own right, but if you want to know the structure of a proton, it's better to use electrons since they have no inner structure." For Herwig's new research interests, that dictated an electron machine, but the advantages did not stop there. "A circular electron machine is also better to do such measurements because you can use an internal target that is installed inside the magnet ring of the accelerator itself," he continued. "If you want to investigate the structure of the proton, you can't bombard a single proton, you use a liquid hydrogen target. Protons are the nuclei of hydrogen atoms, so they are packed very densely in liquid hydrogen. And if you want to study all the details, you need many, many electrons to sound out the structure of the proton. Because the proton is so small, most of the electrons you fire at the target pass straight by. In a linear accelerator, you have one shot for each bunch of electrons you accelerate, but in a circular machine with an internal target, if the electrons miss the protons the first time round, they come back again and again and again. You can get more precise results more easily."

One experiment in particular attracted Herwig's attention. Robert Hofstadter had pioneered the technique for investigating the structure of nuclei and individual nucleons (protons and neutrons) by bombarding them with high-energy electrons. It was to win him a half share of the 1961 Nobel Prize in Physics, the citation for which reads: "The experimental method used by Hofstadter is connected with the principles of the ordinary electron microscope. Here the possibilities to observe details are increased by raising the voltage which accelerates the electrons." The machine Hofstadter had used for his pioneering work was at Stanford—the two-mile long machine that was on the cards and had a strong pedigree based on smaller linear accelerators going back to the 1940s.

"By measuring the deflection of the electrons from the proton," explained Herwig, "he determined the radius of the protons and also of heavier atomic nuclei for the first time, and by scattering electrons from deuterons, which consist of one proton and one neutron, he could also measure the radius of neutrons." Herwig wanted to take this research a step further, and he recognised that with the Cornell synchrotron and its internal target, he was in exactly the right place to do so.

"The proton carries a positive electric charge that is not concentrated in a point but has a local distribution characterised by an average radius," said Herwig. "But although the overall charge of the proton is positive it also contains some negative charges." The same is true for the neutron whose total charge is zero because the positive and negative charges compensate each other. If electric charges are not at rest but move around they produce magnetic fields characterised by magnetic moments that are again not concentrated in a point but have a local distribution. What Herwig wanted to do was to scatter electrons from protons or neutrons, a process through which they would be deflected by the electric charges and also by the magnetic fields, in order to measure the sizes of the particles and the distribution of electric charge and magnetic moment inside them—the so-called proton and neutron form factors.

"The problem is that protons can be used directly as a target but there exist no free stable neutrons," explained Herwig, "so the best one can do is to use deuterium which consists of nuclei containing one proton and one neutron. If one has measured first the properties of the proton one can then extract the properties of the neutron from the measurements on deuterium, but evaluating the experimental data is somewhat complicated."

"In order to get sufficient data, one has to observe scattered electrons at many different angles," continued Herwig. "This is where the advantages of a circular accelerator with an internal target come in." Cornell's electron synchrotron would make it easy for him to collect data from millions of interactions between the electron beam and the protons and deuterons in the target. His next challenge was to build a spectrometer to measure the energy of the scattered electrons. "There was a difficulty," he realised. "If you have an internal target, access is limited because the electrons are kept in orbit by magnets, and they limit the space around an internal target, so we had to invent a new kind of spectrometer that could be installed close to the internal target in a circular machine."

Bob Wilson gave Herwig the task of constructing such a spectrometer, and he hit on the idea of using a quadrupole magnet. The kind of magnets we're most familiar with have two poles, labelled north and south, but electromagnets can be made with as many north and south poles as you can fit into the space available. Quadrupoles have two of each, and were already in use in accelerators for focusing the beams. "So I designed such a quadrupole spectrometer," he said, "It was built at the main mechanical workshop of the institute, and it worked very well for studying electron scattering from an internal target."

With the large amount of data they collected, Herwig and the Cornell team were able to make good measurements of the proton and neutron radii and form factors. "I did something that I am still a little bit proud of: I corrected a mistake that the people had done at Stanford," smiled Herwig. "I showed that although the measurements at SLAC were right, the determination of the electric and the magnetic form factors at SLAC were wrong, so we determined the electric and magnetic form factors of the proton and the neutron in a proper way, and we presented our results at meetings of the American Physical Society. But I must say Cornell was not very good in public relations, unlike Stanford, so I think Cornell never got the credit they deserved for this work."

Another skill that Herwig picked up at Cornell was running both the experiment and the accelerator at the same time. Unlike today's big research facilities where accelerator physics and particle physics are distinct disciplines, and the control rooms for accelerators and experiments are far apart, Cornell's accelerator had been built by the physicists who used it, and it was natural for the control rooms to be adjacent to each other. "We had to operate two things at the same time: the accelerator and the experiment. The control rooms for the two were very close together, only a few metres apart, so we were running back and forth from the control desk of the accelerator to that of the experiment, and operating both at the same time. It was beautiful because you had full control. It gave me the chance to learn how to operate a complicated machine like a synchrotron."

Bob Wilson also provided Herwig with a role model for a hands-on laboratory director. "I remember one remarkable event," he recounted. "One night we realised that we needed a measurement at a large angle of more than 90°, but we could not get the quadrupole spectrometer to cover such an angle because one of the accelerator magnets was in the way, so at three o'clock in the morning, Bob took a hacksaw and cut a corner off the magnet." Herwig took note, but was wary of putting what he'd seen into practice. "Even as Director-General of CERN I would never have dared to touch a machine element without the permission of the expert! Times have changed."

The Schoppers' stay in Ithaca came to an end at the beginning of the summer of 1961, but the family's American adventure wasn't over. "In America, professors and also fellows were paid for nine months of the academic year, and for three months they could do what they wanted," said Herwig. "My fellowship was for the nine months, but we had lived modestly and saved enough money to spend another three months in the United States. I had been invited to give some talks in other universities and laboratories, but for the last six weeks we decided to take a road trip across the United States. I had bought an old Chevrolet, and we went from Ithaca across the great plains and the Yellowstone Park before arriving in San Francisco. I combined this private trip with invitations to laboratories like Los Alamos, where the atomic bomb was developed during the war, and to Stanford. In this way we got to see the beautiful United States, for example driving along the Pacific coast from San Francisco to L.A. We appreciated how easy it was during such a long trip to take care of our baby of less than a year old: in all the motels and restaurants people were extremely accommodating in helping us with Andreas, so we all really enjoyed this trip through the USA, though I don't think Andreas remembers much."

While the Schoppers were enjoying their holiday, political developments in Germany were taking a sinister turn, and would force the family to take a difficult decision: on 13 August, the Berlin wall started to go up. "Bob Wilson was very kind and offered me a job," said Herwig. "Nobody knew what would happen in Germany, even a civil war was not excluded. Since I had an immigration visa, there would have been no problem staying, and in time I would have got United States citizenship, but after long discussions with my wife, we decided to go back to Europe. On the one hand we were fascinated by the American way of life, which had many, many advantages. It's a big country, with apparently no limits, it was fantastic. In the States we could travel for weeks using the same language, the same currency, and staying in the same type of hotels. Beautiful. Europe appeared small, and in a way provincial, with all the diversity of different countries and different languages, but we realised the advantages and the beauty of these diversities, and European culture and way of life, so we decided to go back. I went to Karlsruhe and took up my professorship there."

Through the course of three postings, to Stockholm, Cambridge and Cornell, Herwig had completed the transition from optics to particle physics. A total jump in energy of nine orders of magnitude in two steps.

In His Own Words: Learning About the English Way

"My entry to England gave me an interesting lesson in how things worked there in the 1950s. At that time, of course, aeroplanes practically didn't exist, so surface travel was the only way. I took the overnight ferry from Hamburg to Harwich and arrived, I think, at about five o'clock in the morning. The first thing was I had to go to the immigration office, and they asked me what I wanted to do in the UK. I said, "I have an invitation to Cambridge University for a year." The officer replied, "well, if you want to stay for a year, do you have a visa?" I explained that my fellowship was arranged in a hurry, and I didn't have time to apply for a visa, which would have taken several weeks or months, but pointed out that I was allowed to enter as a German for three months as a tourist. Within those three months, I planned to apply for a visa. He said, "you can't do that, if you want to stay for a year, you should have applied for a visa beforehand. I'm sorry, my hands are bound. You have to leave the United Kingdom on the same boat you came on." I asked him whether he would have allowed me to enter if I had told him that I wanted to stay for three months. He said, "yes, of course, but you told me you wanted to stay for a year." I'd have been better off not telling the truth, but now I was stuck. The boat was scheduled to stay in Harwich the whole day, and sail back to Hamburg overnight. The Immigration officer called a policeman to escort me back to the boat and make sure I didn't try to escape.

So I had breakfast with a policeman, and I really felt like a criminal. But then, at nine o'clock, I learned that there was a change of immigration officer, so I asked the policeman if he'd let me try again, and he took me back to the immigration office, but I got the same answer: "no, our hands are bound. You have to leave the United Kingdom on the same boat you arrived." I asked him why, and he explained it to me. I really only understood the full implication of what he said much later, but it really was an interesting lesson. He told me that since England was never occupied by Napoleon, the Napoleonic Code was never introduced there. The Napoleonic Code meant that law was written down, law was executed according to the written laws. In England, not having been occupied by Napoleon, the legal courts had to refer to previous cases and not to written law. So the immigration officer was afraid that by letting me enter the United Kingdom, he would create a precedent and all German physicists could turn up, refer to the Schopper case, and they would have to be let in. He didn't want to take that responsibility, which I can understand, but I had one last try anyway. I said, "Professor Frisch at Cambridge is waiting for me, and he will be surprised if I don't arrive. Can I at least make a telephone call and inform him about my situation?" He said, "yes, of course, that you can do." So the policeman took me to the nearest telephone booth and I tried to call Frisch. I got his secretary, who told me that he was giving a lecture and couldn't talk to me, but she promised to inform him as soon as possible.

I went back to the boat, still accompanied by the policeman, to wait. We had lunch together, and then, at three o'clock, the immigration officer came in and said, "you are allowed to enter the United Kingdom." I was baffled, and I asked what had happened to make him change his mind. He explained. "Well, Professor Frisch has called us, and we learned that he is a fellow of the Royal Society and that changed the situation." That would have been completely impossible in a European continental country: no scientific society would have had sufficient reputation to change such a decision!

So I was allowed to enter the United Kingdom, and within three months I applied for a visa. It was still a rather bureaucratic process, but I got a visa for the whole year. That immigration officer had given me a very interesting lecture concerning the different way that laws in Europe were applied, and it wasn't the last lesson I learned about English society that year.

At Cambridge University, I learned that the courtyards of the colleges are, in principle, private, but for many centuries the practice has been that the public can walk through them. In order to keep that up, however, it is necessary that, at least

Fig. 5.7 Otto Frisch pictured at Cambridge around 1970 with Sweepnik, a device he designed to analyse bubble chamber images (Courtesy of Cavendish Laboratory, University of Cambridge, © University of Cambridge, All rights reserved)

once a year, two members of the public have to walk through the courtyards and give neutral testimony to each other. These old institutions are full of stories like that, which seem quaint, but I liked them. Another that amused me concerned a penalty given to a student for a rather unusual misdemeanour. At that time, all the students and scientists had to wear gowns. I also had to wear a black gown. One day they caught a student swimming naked in the River Cam, he was called before a provost and he was told that it's forbidden to swim naked in the Cam and they wanted him to pay a few pounds as a penalty, but they couldn't find a precedent case where a student was punished for swimming in the Cam. In the end, he was punished for not wearing his gown while swimming in the Cam, and they managed to fine him, so that was another nice lesson in English law.

My learning curve was not confined to the university. When my wife and daughter arrived, we had difficulties finding an apartment. We didn't have much money, so I couldn't afford a house, and we ended up renting an apartment in an old house. The landlady was very nice, and she lived on the ground floor. We were on the first floor. The house didn't have central heating and winter was coming. There was a gas fire, but to keep it going, you had to feed it with a shilling every two hours, I think it was:

a shilling to keep the room warm. So we had to learn to get up during the night to put in a shilling into the heater, so as not to freeze to death. There was no hot running water, and once a week, on Saturday, the water was heated and the whole family could have a hot bath.

Our landlady had children of her own, and she included my daughter in their games. We were happy there, but again I learned something about the English mentality. One day, she asked me what I did. I said, "I'm with Professor Frisch at the university. I thought she would be impressed that he was a professor there, and a member of the Royal Society. But instead, she observed "Oh, he's a German immigrant." When I explained to her that Frisch was the nephew of Lise Meitner, that he'd had to leave Europe as a refugee, just as Meitner had gone to Sweden, and he had been in England for more than 20 years, fully recognised in academic circles, she only said, "no, no, a pear is a pear, and an apple is an apple. He's not British, he will always remain an apple and never become a pear." She didn't mean any harm, but that was a lesson in English mentality: not in academic circles, but among ordinary people. Later, we found a more modern apartment, with central heating and hot water."

Reference

1. Schopper H (1957) Circular polarization of γ-rays: further proof for parity failure in β decay. Philos Mag J Theoret Exp Appl Phys 2(17):710–713

Chapter 6
To DESY via CERN

In 1970, the Schoppers moved to Geneva. Herwig had been offered a position as head of the laboratory's Nuclear Physics Division, and he took unpaid leave of absence from Karlsruhe to take up the post. It was a timely move, since following his first stint at CERN, he'd set up a CERN user group at Karlsruhe, doing experiments at the proton synchrotron (PS) and later the Intersecting Storage Rings (ISR). The group would later benefit from CERN's agreement with the Soviet Union, signed in 1967, to go to the world's highest-energy accelerator of its day at Protvino, near Moscow.

A Tale of Two Machines

Herwig's second stay at CERN coincided with a curious, yet vitally important period in the laboratory's history. By the mid-1960s, there were competing ideas as to what CERN's next major facility should be. The engineers were pushing for a proton collider, the ISR, which would be able to generate collision energies much higher than a fixed-target machine like the PS. But the physicists were worried that a collider would not produce enough collisions to do meaningful physics. Nobody had ever built one before, and to some it looked like a risky route to follow. They wanted to play safe, and were advocating a larger version of the PS, a super proton synchrotron, instead. By 1964, however, CERN's Director-General, Viki Weisskopf, was convinced that a collider would work, and he set about persuading the CERN council to approve the project, while also presenting parallel plans for a 300 GeV proton synchrotron. His strategy was to get both machines approved, with the ISR coming first.

It was a strategy that paid off, and the long-term consequences are still playing out. The ISR came on-stream as the world's first hadron collider in January 1971, with the Super Proton Synchrotron (SPS) following five years later. By this time, there was no doubt that a hadron collider could provide enough collisions for physics: working with the ISR had taught CERN's scientists and users much about operating such a

© The Author(s) 2024
H. Schopper and J. Gillies, *Herwig Schopper*, Springer Biographies,
https://doi.org/10.1007/978-3-031-51042-7_6

Fig. 6.1 Herwig giving an interview on succeeding Peter Preiswerk as Head of CERN's Nuclear Physics Division in 1970 (©CERN, All rights reserved)

machine, and building collider detectors. In 1976, just as the SPS started running, one such scientist made a bold proposal to convert it into a collider.

Carlo Rubbia argued that by doing so, experiments would be able to provide definitive evidence for a theory developed in the 1960s by Sheldon Glashow, Abdus Salam and Steven Weinberg that brought the electromagnetic and weak forces of nature together in a single theoretical framework. The electroweak theory required the existence of heavy force-carrying particles, labelled W and Z, and although evidence had been found for the Z by the Gargamelle experiment at the PS in 1973, the particles themselves remained to be discovered.

CERN's big project for the future was to build a large electron–positron collider (LEP) as a W and Z factory that would put electroweak theory to the test, but Rubbia saw a way to find the W and Z particles sooner. Converting the SPS to a collider was not trivial, but it was nevertheless easier than building a collider from scratch.

As a proton–proton collider, the ISR consisted of two rings, as its name suggests, so that two beams of protons could be made to circulate in opposite directions and collide at the intersection points. To turn the SPS into a collider, a different approach would be needed, as the SPS is just a single ring. The solution was to turn it into a proton–antiproton collider in which protons and antiprotons could be made to circulate in opposite directions inside the same magnetic ring. That required the construction of a facility to make and store antiprotons in sufficient quantities to make a beam.

In 1981, the SPS started to run as a collider, with Rubbia at the helm of one of its two experiments, known as UA1. Two years later, the discovery of W and Z particles was announced by the SPS collider's experiments, and the following year, Rubbia collected the Nobel Prize for Physics along with engineer Simon van der Meer, who had made hadron colliders possible through his development of a technique used to collect and marshal beams of antiprotons. The ISR was decommissioned in 1983, but it had changed the course of physics forever.

An Offer Too Good to Refuse—Back to Hamburg

When Herwig arrived at CERN in 1970 it was the year before the ISR started up, and SPS construction was well underway, coordinated from a new laboratory in France, just a few kilometres from the original CERN site in Switzerland. At the end of the year, CERN Director-General Bernard Gregory's mandate came to an end, and he was replaced by two Directors-General, Willibald Jentschke, who had been recruited from DESY, to run the original CERN laboratory, which was now known as Lab I, with John Adams in charge of Lab II and the SPS.

Jentschke's move meant that there was a vacancy at DESY, and in 1973, Herwig was invited to fill it. "I received a letter from Hamburg, and I remember it was a weekend," recalled Herwig. "My son had just got back from skiing in the French Alps and he laughed when I told him. He'd been up at 3000 metres, and the highest hill in Hamburg is just 100 m above sea level. There's a big difference between Geneva and the plains of northern Germany." Nevertheless, the offer was too good to refuse, and the Schoppers were soon on the move. Part of Jentschke's legacy was that as a national laboratory, DESY was independent of the university, so the post Herwig was offered was a dual one: the chair of the DESY directorate and a full professorship at the university. It was an arrangement that was right up his street.

Unorthodox housing arrangements were a trademark of Schopper family moves and the move to Hamburg was no different. The government of Hamburg had offered them a parcel of land, but there was no house on it, so while one was being built, they moved into the DESY guesthouse. "I rented two apartments in the guest house on different floors, and our dog, a collie, had to learn to move from one to the other one by taking the elevator. Another problem was that by this time we had a lot of furniture. I had a grand piano, and these apartments were too small." Luckily for the Schoppers, DESY had a lot of storage space, and another Jentschke legacy even provided an ideal solution for the piano. "Jenschke had insisted on having a very large office, and there was enough space in it for my grand piano. For my first two years at DESY I had an office with a grand piano and I could play it in the evenings."

Fig. 6.2 Herwig attends a lecture in the CERN Auditorium in 1971. Director-General Bernard Gregory is seated two places to his right (©CERN, All rights reserved)

DORIS: A Collider, not a Girl

Herwig took over from Wolfgang Paul, who had stepped in on a temporary basis when Jentschke had left. He inherited an ambitious project that had been initiated by Jentschke in the 1960s: an electron–positron collider called DORIS, which stands for *Doppel-Ring-Speicher* (double-ring store). With Europe's big laboratory CERN following an exclusively proton route and building the ISR and SPS, Jentschke had chosen a complementary path for Germany's new national laboratory. "DORIS was a collider system of two rings," Herwig explained, "the idea was that it could collide electrons with electrons. Most lepton colliders collide negatively charged electrons with their antiparticles, positrons, which have a positive electric charge. Because of their opposite charges the two kinds of particles can orbit in the same magnetic ring in opposite directions, but the downside is that it's hard to get strong positron currents since the positrons have to be created in a separate machine. DORIS's two rings avoided this problem and as well as being able to generate a high number of collisions leading to higher precision results, the expectation was that DORIS would be able to get different information than colliding electrons with positrons.

Fig. 6.3 Five Chairs of the DESY directorate. From left to right, Herwig Schopper (1973–1980), Wolfgang Paul (1971–1972), Willilbald Jentschke (1959–1970), Volker Soergel (1981–1993) and Bjørn Wiik (1993–1999) (©DESY, All rights reserved)

In theoretical terms, in electron–electron collisions the normal Coulomb interaction is at play, whereas in electron–positron collisions the particles can also interact by annihilating each other and then producing new particles. This allows interactions due to the electromagnetic force to be separated from those due to the weak force. In the end, however, since DORIS produced enough collisions in electron–positron mode, it was operated mostly in this way."

Herwig recounts a tale that shows how far the field has come since the early days of colliders. "When Jenschke started to discuss the idea for DORIS, he asked all the big theoretical physicists at DESY like Harry Lehmann and Kurt Symanzik what would be the highest useful energy for an electron–electron or electron–positron collider," he recalled. "Believe it or not, their answer was 2 GeV." The theorists based their argument on the fact that the cross-section for the collision of point-like particles, such as electrons and positrons, goes down with the square of the collision energy, so the higher the energy, the lower the number of collisions. In addition, all form factors for the production of non-point-like particles are smaller than one and so they thought that above 2 GeV, there would be so few collisions as to make physics research impossible. "But Jentschke," said Herwig, "being an experienced experimentalist, thought that nature would be more imaginative than the theorists, and planned for a higher energy machine. In an electron storage ring, the magnet system is not the most costly system, it is rather the high-frequency system that accelerates the particles. The reason is that the highest energy achievable increases linearly with the radius of the machine in a way that is not so dramatic, but the losses due to the emitted synchrotron radiation go up with the fourth power of the energy."

Jentschke decided to go for an energy of 5 GeV, modest by today's standards—the highest energy electron–positron collider to date was CERN's LEP, which accelerated beams to over 100 GeV, and was built while Herwig was Director-General of CERN in the 1980s—but Jentschke's choice turned out to have important consequences for DESY, giving access to a range of new physics as the years went on. "It was a very wise decision," said Herwig, "it shows you should not always take the advice of experts at face value."

DORIS started up in 1974, the year that Herwig arrived, with a beam energy of 3.5 GeV, later upgraded to 5 GeV in 1978. There were two particle physics experiments, called PLUTO and DASP. A number of other experiments also made use of the synchrotron radiation that sapped the energy of the beams, but provided a rich seam of research potential in fields other than particle physics. DORIS would run until 2013, and for much of its life, it was used exclusively as a synchrotron radiation facility for experiments ranging from solid state physics to life science.

The Discovery of Charm

"At the beginning nothing very dramatic happened, nothing very surprising was found," said Herwig. "Then something completely unexpected happened. There was a young physicist, Samuel Ting, who later became a professor at MIT. He had worked at DESY for several years in the 60s, and had a good feeling for where new physics could be discovered. At that time there were theories predicting new kinds of particles, which we now call charm particles. These theories could predict most of the properties of such particles, but not their masses. Since it is the mass of a particle that determines the minimum energy required to produce it, according to Einstein's famous equation $E = mc^2$, that made them rather hard to look for. We had no idea how heavy these new particles would be, which meant that to look for them you would have to scan through a wide range of energy, which is rather tedious with an electron–positron or electron–electron collider. Ting realised that he could do an experiment covering a wide band of energy with a proton machine, so he decided to spend some time at the Brookhaven National Laboratory in the US where they had a proton machine with sufficient energy to look for these charm particles, and indeed he found them there. Almost at the same time, they were detected at Stanford by a team led by Burton Richter."

This discovery became known as the November revolution in physics, because it triggered a chain of events that would reshape our view of the structure of matter, and set the direction for particle physics for many years to come. The teams led by Ting at Brookhaven and by Richter at Stanford had discovered a particle that is unique in that it had a double-barrelled name, the J/Psi. J resembles the Chinese character for Ting's name, while Psi was the name given by the Stanford Linear Accelerator

Centre (SLAC) team, because tracks left by the decaying particle resemble the Greek character. "Ting's and Richter's papers were published in the same journal side by side," said Herwig. "However, there were rumours that Ting's first results leaked out, giving Richter a hint at which energy to look at, which would have made things easier since Stanford would then have had to scan only a small energy region. There's no doubt that both deserved the Nobel Prize since such a fundamental discovery needs a confirmation by two independent experiments."

The J/Psi is a meson, a particle consisting of a quark and an antiquark. Before its discovery, only three types of quarks were known: up quarks and down quarks, which group together in threes to make up the protons and neutrons of ordinary matter, and strange quarks, which had made their presence known in the cosmic radiation constantly bombarding us from space. Theorists had been speculating about the existence of a fourth quark since the 1960s, but in 1970 Sheldon Glashow, John Iliopoulos and Luciano Maiani published a compelling paper that required the existence of a such a particle.

The J/Psi consists of a charm quark and an anticharm quark, and its discovery was effectively the experimental foundation stone of the Standard Model of particle physics. It won the Nobel Prize for Richter and Ting in 1976—this time, the experimentalists reaped the reward, while the theorists who predicted the new particle went empty handed, although Glashow would go on to receive the honour three years later.

Particle physics abounds with stories of scientists who could have discovered the J/Psi if only they had been looking in the right place. Apart from the machines at Brookhaven and Stanford, there were several others in the world that had the required energy, including DORIS at DESY and the ISR at CERN. "CERN could have detected the J/Psi but the guidance by theory was too limited," said Herwig. "At CERN it was mainly collisions in the forward direction that were considered to be of interest, but to detect a new particle produced at its threshold energy, you need to look for particles coming out at large angles. In the first round of experiments at the ISR, nobody had a detector at large angles: they were only installed in the second generation and they immediately saw the J/Psi and the gluon. It was a pity that these were missed. Lesson: do not believe too much in theory."

Once the discovery had been made, however, experiments at DORIS started to study the new particle, and it was then that Jentschke's hunch to overrule his theorists really paid off. Although proton machines are good for discovery, electron machines are typically the machines of choice for precision physics, and PLUTO and DASP were in for a rich harvest, competing, and collaborating closely with Richter's team at Stanford. "The mass of this new particle was just in the mass range accessible by the DORIS ring," said Herwig, "so when we learned about that, the DORIS experiments immediately jumped on it and started to look for the J/Psi. During the first years I spent at DESY, the most interesting experiments were done with DORIS when they were investigating this newly discovered family of charm particles."

The J/Psi particle is a charm–anticharm pair with parallel spins adding up to a total spin of one, but the theory predicted that charm quarks should also pair with opposite spins giving a total spin of zero. "The first major success of the DORIS experiments was that they could discover a new particle of the charm family, the so-called η_c (eta sub-c) particle, along with other charm-quark-containing mesons in which the charm quarks are paired with other kinds of quark and antiquark," explained Herwig. "These were very fruitful years, and there was healthy and friendly competition with SLAC, with a lot of data being confirmed mutually. To make the quark model credible, it was necessary to have consistent experimental results from several experiments."

Charm physics turned out to be just the start for DESY, with DORIS putting Germany's national lab for particle physics firmly on the international map. However, it was not long before another group of physicists in the US, this time led by Leon Lederman at Fermilab, was to discover a fifth type of quark. First predicted in 1973 by Japanese theorists, Makoto Kobayashi and Toshihide Maskawa, the bottom quark made its appearance in the form of the upsilon, discovered in 1977.

A playful sense of humour seems to surround everything that Lederman did. When his book about the Higgs boson, The God Particle, appeared, he said the title was his publisher's idea. He'd wanted to call it the Goddam Particle because it was so hard to find. Lederman had originally announced the discovery of the upsilon in 1976, but the announcement turned out to be premature—what had appeared to be a new particle was just an illusion, and thereafter, the initial discovery came to be known as the Oops-Leon, leaving upsilon for the real particle.

By the time the bottom quark made its appearance at Fermilab, DORIS was running at 5 GeV per beam, enough to produce upsilons abundantly. This allowed the DORIS experiments to clarify Lederman's discovery once and for all. "The mass resolution at Fermilab was bad," explained Herwig. "They discovered a peak that was relatively wide, so there was a suspicion it was not one particle but several with similar masses. A few weeks after the discovery at Fermilab, DORIS showed that the big peak at Fermilab was not one particle but in reality, three with slightly different masses that DORIS could resolve beautifully. DORIS showed for the first time that there was a whole family of B particles, as predicted."

Another Broken Symmetry—CP

Kobayashi and Maskawa had predicted the bottom quark in a theoretical answer to an experimental conundrum. Back in the 1950s the world had been introduced to the importance of symmetry, and in particular the way that nature can subtly break symmetry, by Lee and Yang when they proposed parity violation in beta decays, as well as by Salam, as Herwig discovered at Harwell in 1956 (see Chap. 5). When Herwig learned this, it changed the course of his career, leading him to publish one of the first experimental verifications of parity violation.

The notion of symmetry has profound consequences for our understanding of nature, with broken symmetries turning out to be one of the most important things in shaping the universe as we know it. The drama of parity violation, denoted by the letter P, had revealed that particle–antiparticle symmetry, denoted by a C, for charge conjugation, was also broken. However, the combination of charge and parity symmetries, CP, was considered to be sacrosanct. In other words, any particle inter-action should be indistinguishable from the mirror image of that interaction in which the particles are also swapped for their antiparticles. While each symmetry could be broken individually, together they would be strictly conserved.

Another surprise was, nevertheless, waiting. Although not predicted by theory, it soon turned out that the combined CP symmetry is also violated, although not to the maximum degree possible, like C and P separately, but just a little bit. When this was demonstrated in 1964 by James Cronin and Val Fitch, it really upset the applecart of fundamental physics: it was a full decade before Kobayashi and Maskawa proposed their viable explanation for CP violation, along with the prediction of the bottom quark.

This time, both the theorists and the experimentalists were recognised with the award of a Nobel Prize. Cronin and Fitch received theirs in 1980, with Kobayashi and Maskawa having to wait until 2008 for the honour. As for Lederman, he received a Nobel Prize in 1988, but for work conducted earlier into neutrino physics.

DORIS's Last Particle Physics Hurrah!

DORIS's successes emboldened Herwig to plan for a bigger machine at DESY. It would be another electron–positron collider complementary to the big proton machines at CERN, and it would be known as PETRA. When it came on stream in 1978, most of the particle physicists moved to the more powerful machine, but DORIS still had one more particle physics trick up its sleeve. "I thought that it would be a pity not to continue to use DORIS for particle physics, so I convinced one of my former Ph.D. students from Karlsruhe, Walter Schmidt-Parzefall, to start a new collaboration and propose a new detector for DORIS. I promised him my full support as DESY director and after some hesitation he agreed. He managed to bring together a group of physicists and against most people's expectations this late experiment at DORIS produced a great result." Schmidt-Parzefall's collaboration consisted of groups from Russia, Germany, the United States and Sweden, giving rise to the acronym, ARGUS, and they were later joined by groups from Canada and Yugoslavia. Construction began in 1979, and the experiment ran from 1982 to 1992. The big result came in 1987. "ARGUS was the first place where the conversion of a B-meson into its antiparticle, an anti-B-meson, was observed," explained Herwig. "From these so-called B-oscillations one could conclude that it was possible to convert the bottom quark into a different quark. This was one of the most important results obtained at DESY and it came from a facility that was considered as obsolete by many scientists. From this ARGUS data one could also conclude that the as yet undiscovered sixth quark—the top quark—had to possess a huge mass, much higher than previously thought." The top was eventually discovered at Fermilab in 1995, with a mass of over 172 GeV/c^2, about 170 times heavier than a proton.

From DORIS to PETRA

Herwig was not alone in his ambition to build a big electron–positron collider. In the UK, the Rutherford Laboratory had plans for a similar machine, and at Stanford, there were plans to build bending arcs at the end of the two-mile Stanford linear accelerator to turn it into an electron–positron collider. Herwig was fortunate that his predecessor, Wolfgang Paul, had hired one of the foremost accelerator builders of his time, Gustav-Adolf Voss, who'd been working in the States on the Cambridge electron accelerator, a Harvard/MIT project that evolved under his guidance into a colliding beam electron–positron storage ring. Voss would go on to play a significant role at a later stage in Herwig's life, as a pioneer of the SESAME Laboratory in Jordan, of which Herwig was the first president.

"With him we discussed what the next machine at DESY could be," recalled Herwig, "and after some discussions, and taking into account the possible cost and the available size of the site, we came up with a proposal to build an electron–positron machine with a maximum energy of about 20 GeV." If successful, it would be the largest facility of its kind on the world. They named the project the *Positron-Elektron-Tandem-Ring-Anlage* (positron–electron tandem ring facility), PETRA. "I decided to continue the tradition of giving DESY's machines female names," recalled Herwig, "a sign of the times, I suppose, since PETRA's positron injector, the Positron Injector Accelerator (PIA), followed suit. It was not easy to get PETRA approved, but I was optimistic after the success of DORIS."

Fig. 6.4   Chancellor Helmut Schmidt visited DESY on 11 March 1977 to take stock of progress on the PETRA storage ring (©DESY, All rights reserved)

Fig. 6.5 Herwig raises a toast on 27 January 1976 with German research and technology minister Hans Matthöfer on the occasion of the laying of the foundation stone for PETRA (©DESY, All rights reserved)

Synchrotron Radiation—A Valuable Spin-Off

While the approval process for PETRA was ongoing, and international discussions got underway to decide which facilities would get built, and which countries would support which others, DORIS set the direction for the laboratory that DESY would eventually evolve into. From the very beginning DESY had been a laboratory with a dual mission: fundamental physics using the synchrotron's particle beams, and a range of scientific applications using the so-called synchrotron radiation given off by the circulating beams.

"From the very beginning at DESY, the synchrotron radiation, which is a nuisance for particle physics experiments, was recognised as an important research tool in its own right," explained Herwig, "circulating electrons emit radiation with wavelengths ranging from infrared to hard x-rays, and several beamlines for this synchrotron light were installed at DORIS for solid state physics and biology experiments."

With the arrival of PETRA, DORIS's main focus was as a synchrotron light source, and in 1980 a special laboratory, HASYLAB, with more than 19 beamlines was opened for German and international researchers. "Christoph Kunz, Ernst Koch and Ruprecht Haensel, who later became director of the European Synchrotron Radiation Facility in Grenoble were among the key people to promote synchrotron radiation science at DESY," recalled Herwig. "I also signed an agreement with the Director-General of the European Molecular Biology Organization, John Kendrew, to install the organisation's first outstation at DESY." Over time, DESY's main focus would evolve from particle physics to synchrotron light source science, with PETRA itself becoming one of the world's most prominent synchrotron light sources and a powerful example of technology developed for particle physics serving other disciplines of science.

This, however, was not the only time that Kendrew's path would cross with that of Herwig, and the second time the circumstances were not so happy. A decade on, science funding in the UK was under considerable pressure. A committee was established to examine the country's involvement in particle physics in general, and in CERN in particular. The now-enabled Sir John Kendrew was the chair. The committee recommended a 25% reduction in the UK's contribution to the laboratory, or if that proved not to be possible, complete withdrawal from CERN. This all happened in the middle of Herwig's tenure as Director-General (see Chap. 7).

The Electron Collider Race to 20 GeV

The course of true love never did run smooth, and so it was for the partnership that emerged between the Rutherford Laboratory and DESY over negotiations to build a 20 GeV electron–positron collider. "I must say I was under strong psychological pressure because first the British said that they urgently needed a new facility at the Rutherford lab," recalled Herwig, "but even more so because of SLAC."

The Rutherford Laboratory's 7 GeV proton synchrotron, Nimrod, which had started up in the early 1960s, and included a verification of Cronin and Fitch's observation of CP violation among its accolades, was reaching the end of its life. At Stanford, Pief Panofsky, the founding director of SLAC, was still at the helm. "Panofsky and I we were friends, we talked very often together," recalled Herwig. "He told me, 'Look, you are crazy to try to do something like that at DESY because we will do it faster and better—we have already the LINAC accelerator, it's much easier for us to add these two half circles and get collisions. You have to build a new tunnel, and we have much more experience than you, so our machine will be finished before your machine, and we'll do all the exiting physics before you can.' My answer was to tell my friend that we'd accept the challenge and compete."

PETRA required approval from both the federal government and the State of Hamburg. "I managed to get approval in an incredibly short time, a matter of a few months, thanks to the Federal Research Minister at that time," said Herwig. "His name was Hans Matthöfer, and I must say I have met and worked with many research ministers in my life, both in Germany and other countries, but he was one of the best. Although, and perhaps because he was not a scientist, he was prepared to listen to the advice of scientists, and he developed a good feeling for the quality of proposals and the people who presented them. In addition, he came from the trade union, he was a union man, and that gave him a very strong position in the Federal Cabinet to push through decisions and get PETRA approved."

Fig. 6.6 When PETRA was inaugurated in 1979, Germany's President, Walter Scheel (front) was in attendance, accompanied by research and technology minister, Volker Hauff, who had succeeded Matthöfer in 1978 (©DESY, All rights reserved)

Matthöher belonged to the Social Democratic Party, and the fact that the State of Hamburg also had a Social Democratic government at the time did no harm to Herwig's cause. "The Federal Government felt an obligation to give special support the North German regions, which were a little bit neglected at the time compared to South German states like Bavaria."

Federal and state-level approval was necessary for PETRA to go ahead, but was not in itself sufficient. Herwig also needed the approval of the next layer of hierarchy. For DESY, that meant that the *Verwaltungsrat*, an administrative council whose members included representatives of the federal government and the Hanseatic City of Hamburg, also had to deliver a formal endorsement. "For many years the *Verwaltungsrat* was chaired by Günter Lehr, and Hermann Schunck," recalled Herwig. "Josef Rembser, who later became President of the CERN Council, was also involved. The three were consecutive Directors-General in the Ministry for Research, and made very important contributions to science, in particular to particle physics."

With PETRA approved in the autumn of 1976, the case for the Rutherford Laboratory's new facility was greatly weakened, and Nimrod was destined to be the last major domestic facility for particle physics in the UK. Although the laboratory still hosts a thriving particle physics community, its major on-site facilities are now in the area of lasers, light sources and spallation neutron science.

"To provide some consolation to our British colleagues, we invited them to visit DESY, and I proposed that we should establish a collaboration," said Herwig. "One evening we had dinner in one of the restaurants near the River Elbe, and over dinner I explained that the room we were dining in was usually used for marriage parties, so I said, 'I hope our meeting here will also symbolise a union between DESY and the Rutherford Laboratory,' but the director of the Rutherford Laboratory answered in somewhat bitter terms, so the marriage, unfortunately, was not to be." The British were still clearly smarting at the loss of what might have been.

PETRA was soon up and running. "Thanks mainly to Gus Voss and his colleagues, PETRA was built in a record time of two years and eight months, with 20% less budget than planned, and it came into operation in the autumn of 1978," said Herwig, "almost two years ahead of our competitors at SLAC." Herwig had beaten his old friend Panofsky.

Nimrod switched off in 1979, as the UK focused its efforts on facilities overseas, contributing to the programmes at CERN, SLAC, and DESY, while the Rutherford Laboratory explored pastures new. There may have been no marriage between DESY and the Rutherford Laboratory in the 1970s, but what emerged was a strong and successful international partnership between particle physics laboratories around the world. Competition was fierce, but collaboration more so: a hallmark of modern particle physics.

Physics at PETRA and the Discovery of the Gluon

The approval of PETRA put DESY firmly on course to becoming an international laboratory. Although international collaborations were already well-established at DESY, notably with the ARGUS detector at DORIS and the European Molecular Biology Organization's outpost, it was at PETRA that they firmly took hold. Each of PETRA's four interaction points hosted an experiment run by an international collaboration. "Four detectors were approved for PETRA in 1977 after the usual extensive discussions and evaluations in the appropriate international advisory committee," said Herwig. "They were all proposed and constructed by international teams of outstanding scientists." The experiments were named JADE, Mark J, CELLO and TASSO. JADE was a simple concatenation of Japan—Deutschland—England. Mark J was an experiment headed by Sam Ting, who returned to DESY after his Nobel-winning exploits at Brookhaven. The CELLO collaboration was a mainly Franco-German collaboration, while TASSO, the Two-Arm Spectrometer Solenoid, was led by the Norwegian Bjørn Wiik, who would later become DESY's director. When these four experiments were approved, it was agreed that PLUTO, after one last run at DORIS, could occupy one of PETRA's interaction points until the last of the four new detectors was ready to be installed.

PETRA collided its first beams in September 1978, with data taking starting in earnest in January of the following year at Mark J, PLUTO and TASSO, JADE having suffered damage in a pilot run and CELLO not yet being ready. "There were two main questions that we hoped to solve with PETRA, looking into an energy range where theorists were unable to make many predictions, but where we thought some gold might lie," said Herwig. "One was the search for the top quark, and I remember one day a very well-known German theorist came and told me, 'Look, I have a new theory, and I am sure that the mass of the top quark must be in the range of about 22 GeV.' PETRA was designed for 20 GeV, but the magnets were powerful enough to

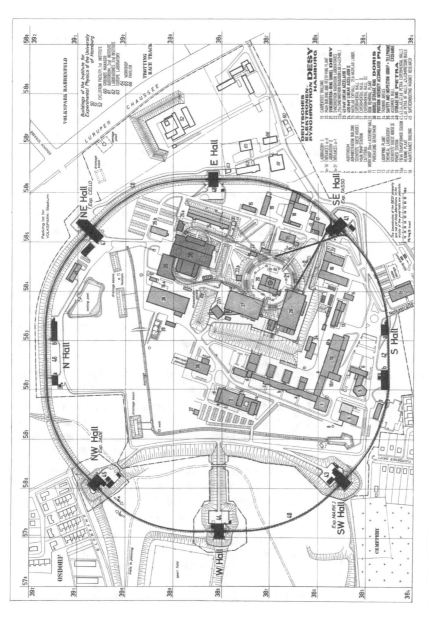

Fig. 6.7 A plan of the DESY laboratory showing the position of the PETRA ring in red (©DESY, All rights reserved)

handle higher energy beams if more accelerating cavities were added, so after some discussions with the committees, we increased the energy to 27 GeV, but there was no top quark to be seen, much to our disappointment. As we know now, the top quark is much, very much, heavier and was discovered many years later at Fermilab."

The second big hope for the PETRA experiments resulted in a major discovery in 1979. "The other burning topic at the time was the search for the carrier of the strong nuclear force, which is called the gluon because it is responsible for binding together the quarks, and consequently the constituents of the atomic nuclei, protons and neutrons, and hence guarantees the stability of matter," explained Herwig. "It seemed difficult to find, but there was a prediction that gluons would decay into three jets of particles, which would be an unmistakable signature for them. The gluon's mass was not fixed and so there was no particular energy range to search, but the consensus was that the best place to look would be at high energy because the individual jets would be easier to identify. The experiments started to look for the gluon."

When an electron and a positron collide, the collision can give rise to a quark and an antiquark, each of which cascades into a jet of particles emerging from the collision. The theoretical prediction that the PETRA experiments were looking for was the occasional emission of a gluon, which would lead to three jets of particles, not just two. In addition, all three jets would be in the same plane. As PETRA's energy was gradually ramped up, from 13 GeV initially then to 17 GeV, and finally by spring 1979 to 27 GeV, all the experiments were keenly scanning the data for three co-planar jets. "Each experiment had a different approach," explained Herwig, "each focusing on different properties of the gluon, but in the end all four found the gluon, with even PLUTO producing interesting gluon data. This, was a great success and a very happy event."

Happy though the particle physics community may have been at this momentous discovery, the question of who got there first was a matter of some contention. "There were some heated discussions, particularly between Mark J and TASSO, whose leaders were Sam Ting and Bjørn Wiik," recalled Herwig. "Both claimed that their experiment had seen the gluon a little bit earlier than the other. I followed these discussions closely, receiving reports almost daily. I think it was a futile fight: only the results of all four experiments together provided convincing evidence since the signatures were not so clear that one experiment alone could claim that they had established the existence of the gluon beyond any doubt." Over time, the community gave the discovery the recognition it deserved, with the European Physical Society devoting two awards to the discovery in 1995. One, the annual High Energy and Particle Physics Prize went to Paul Söding, Björn Wiik, Günther Wolf and Sau Lan Wu from the TASSO collaboration, while the second, a special one-off prize, honoured all four experiments.

"This discovery was a great success for DESY," remembered Herwig, "but later it also produced some disappointment. To understand how something could be so important, a bit contentious, and also disappointing at the same time, it's worth taking some time to understand the physics behind it." In quantum physics, the actions of nature are described by the exchange of force-carrying particles between particles of matter. There are four fundamental forces at work in the universe today, governing everything from the movement of stars and galaxies to the inner workings of atoms. They are gravity, electromagnetism, the weak nuclear force and the strong nuclear force. Gravity governs the large-scale behaviour of the universe, but it is the weakest of the forces by far. No verified quantum theory of gravity exists, and there are no techniques to study it at the particle level. The great success in this domain was the first observation of gravitational waves by the LIGO experiment in 2015.

Particle physics focuses on electromagnetism and the nuclear forces. Among the carriers, the photon is the best known. It not only transmits light but is also responsible for binding electrons to nuclei to form atoms, and atoms to each other to form more complex structures. The existence of the photon was established at the beginning of the last century.

The weak nuclear interaction is necessary to understand nuclear beta decay and is the driving force behind energy production in stars. It is carried by electrically neutral Z particles, a bit like heavy photons, and charged W particles. The discovery of these particles at CERN was announced in 1983. The strong nuclear force, which confines quarks to form protons and neutrons, and binds protons and neutrons into nuclei, is carried by the gluon.

The overall picture of particles and their interactions is called the Standard Model, and it has one more ingredient: the Brout-Englert-Higgs (BEH) mechanism, which accounts for the masses of many of the fundamental particles. Its existence was confirmed at CERN by the 2012 discovery of a particle that carries the name of Peter Higgs.

At the time the gluon was discovered at DESY, this picture was still being painstakingly pieced together, and each new particle discovery provided an important part of the puzzle. "The discovery of particles had been a rich source of Nobel Prizes," explained Herwig, "so there were hopes that this would be the case for the discovery of the gluon, but that didn't happen." When two CERN scientists, Carlo Rubbia and Simon van der Meer, were awarded the Nobel Prize in 1984 for their contributions to the discovery of W and Z particles, the feeling of disappointment at DESY must have been very keenly felt. "The problem was that the gluon was discovered by several experiments involving too many outstanding scientists whereas the Nobel Prize can only be given to three people," said Herwig. "That seems unjust in my opinion because it doesn't do justice to experiments carried out in collaborations. The Nobel Committee's three-person rule is more easily fulfilled by theorists than experimentalists."

In recent physics history, Rubbia and van der Meer have been the exceptions: it's rare for experimentalists to receive the Nobel Prize. The discovery of the gluon was crowned with a Nobel Prize: for the theorists David Gross, Frank Wilczek and David Politzer who pioneered the theory of strong interactions, quantum chromodynamics in the 1970s. They received the call to Stockholm in 2004, curiously one year before the EPS rewarded the DESY experiments with its awards.

When the ATLAS and CMS experiments at CERN announced the long-anticipated discovery of the Higgs particle in 2012, it was a similar story. Two large experiments taking data at the world's most powerful particle accelerator and relying on powerful computing infrastructures were necessary to make the discovery. Thousands of people contributed to the effort, yet the Nobel Prize in 2013 was awarded to two theorists, François Englert and Peter Higgs, Robert Brout having passed away before he could enjoy the experimental confirmation of the BEH mechanism. "There's no doubt that the theorists deserve their Nobel Prizes," said Herwig, "physics requires both experiment and theory, but the three-person policy gives the wrong picture about the nature of research, particularly to young people. I know the Nobel Committee is aware of this problem, but there seems to be no easy solution."

A New Lease of Life for PETRA

The discovery of the gluon was the undoubted highlight of the PETRA particle physics programme, although the PETRA experiments continued to produce good physics up to 1990, when PETRA was repurposed as an injector for a new machine, HERA, the *Hadron-Elektron-Ring-Anlage* (hadron electron ring facility), and as a light source, continuing the tradition established by DORIS. Later, after the shutdown of HERA, PETRA became one of the world's best facilities for synchrotron radiation research. This transition came long after Herwig's watch at DESY and is the source of some amusement to him today. "I was always very much in favour of using the electron machines as sources of synchrotron light, since my original research was not in particle physics or nuclear physics, but was more general so I understood the people wanting to use synchrotron radiation," he explained, "but when I asked the synchrotron radiation people whether they would be interested in using PETRA for their experiments, their answer was: 'Only a stupid high-energy physicist could ask such a question, because if you shot such energetic photons at molecules, they would destroy them.' The characteristic photon energy of the synchrotron radiation with PETRA running at 20 GeV was around 120–150 keV, which was felt to be much too high for atomic or molecular physics. So I did not succeed in attracting any of the synchrotron people to PETRA. That was one of my largest disappointments and surprises at the same time because nowadays, many years later, PETRA has become one of the major and best research instruments in the world, not for particle physics but for biology and solid-state physics because it turned out that PETRA is one of the best machines for synchrotron light if you run it much below its maximum energy. It turns out if a machine has a large circumference like PETRA, the synchrotron light you get is much better concentrated than in a machine with a smaller radius. It's the so-called emittance of the beam that is decisive—the larger the radius, the lower the emittance and the brighter the synchrotron light. After its successful life as a machine for particle physics, PETRA has become a fantastic machine for all kinds of research, and in particular for biology. There's a huge new experimental area for users, and the European Molecular Biology Organization has extended its outpost at DESY. You never know what might become of these old-fashioned facilities."

HERA—A Legacy

Herwig's time at DESY came to an end in 1980, when he received the call to become Director-General of CERN from January 1981. "Before I had the offer to come to CERN, we had discussed a new project to follow the success of PETRA, and this project was the electron–proton collider, HERA, another female name for another DESY machine, continuing the tradition set by DORIS. In the directorate, mostly working with Gus Voss, we had worked out a concept for the project and I presented it to DESY's *Verwaltungsrat* on 6 December 1979. Of course, HERA was realised by my successors, and it was Volker Soergel who promoted it and got it approved. But the concept of HERA was developed when I was still at DESY. So apart from getting

Fig. 6.8 Three Directors-General. Taken at CERN in April 1980, this picture shows Director-General-elect Herwig standing between John Adams, Executive Director-General (left) and Léon van Hove, Research Director-General. The fact that there were two DGs is a legacy of CERN being run as two laboratories at the end of the 1970s (©CERN, All rights reserved)

PETRA approved and built, I am somewhat proud that during my time, DESY was converted from a national facility to an international laboratory. Of course, formally it's still a German laboratory, but I think the users from the outside coming from all over the world feel completely at home there."

In His Own Words

The Chinese at DESY

"One day, I was sitting in my office and the telephone rang. At the other end of the line was Sam Ting. 'Sam,' I said, 'how are you and where are you? Are you in Beijing?' 'I'm sitting in the office of Deng Xiaoping,' came the reply. I said, 'Okay. What is the matter?' He replied, 'Well, I'm discussing with him whether he would be interested to send for the first time Chinese scientists to western countries.' This happened in 1978 soon after the end of the Cultural Revolution at a time when contacts between China and western countries did not exist. So I said, 'Okay. Why not? How many people do you think they would like to send?' He consulted Deng Xiaoping and came back with: 'What about a hundred?' 'A hundred is maybe a little bit too many,' I said. 'Why don't we start with a dozen?' They agreed: 'Okay, we start with a dozen.'

Based on this telephone conversation a collaboration started between China and DESY, and indeed, it was the first time scientists came to a western country from the People's Republic. It was really very moving when the first ten people arrived in April 1979, and they had to learn how to behave in a modern technically developed country. At that time in China there were practically no cars. The main means of moving around was bicycles. In Beijing, the roads were crowded by bicycles, but no cars. So when these people came from the Institute of High Energy Physics (IHEP) in Beijing, they had to learn how to cross the busy roads in Hamburg and not to be runover by cars. Sam Ting had a colleague at that time, Susan Marks, who he later married. Susan became a kind of mother hen to the Chinese group, teaching them how to behave, how to cross the roads, how to do shopping and things like that. It was amusing to see her sometimes walking around DESY or outside being followed by a line of Chinese people like a group of young chicks following their mother. Over the following years physicists came and went. All of them, of course, had been selected before they came to Hamburg. All were excellent scientists. The leader of the first group was Tang Xiaowei and there were also Zhen Zhi Peng and Chen Hesheng, who both became directors of IHEP. Many of those who had spent some time at DESY participated later in the Mark J experiment at DESY, or the L3 experiment at CERN, both of which were headed by Sam. Some went on to outstanding careers going on to fill important posts in science or policy in China. I became friends with some of them and met them quite often later when I visited China.

Later, when I became Director-General at CERN, and Sam Ting followed me to CERN, he proposed the L3 experiment for the LEP machine there. That was the

Fig. 6.9 Herwig with a Chinese delegation at DESY in 1978. Sam Ting is second from right, seated next to his future wife Susan (©DESY, All rights reserved)

first time that Chinese scientists from Taiwan came to a foreign country, and that the Chinese of Taiwan and the People's Republic were allowed to work together in a common experiment. Of course, there was no problem between the physicists, but this collaboration again had to get approval from the highest authorities in Beijing and Taipei. So it's a beautiful example of how fundamental research can contribute to science for peace, creating better relations between governments."

Acknowledgements

"Since DESY was such an important part of my career, I can't end this chapter without acknowledging some of the outstanding people I worked with there. The overall scientific success of DESY depended on not only the original synchrotron and the colliders that followed, but also, of course, the experiments. Many scientists made important contributions and it is impossible to mention them all. The tendency in high energy physics is to emphasise the collaborative effort that is necessary, and names are often not mentioned at all, except in the original scientific publications. I have to follow this tradition also and I can mention only a small number of names, favouring those colleagues with whom I had regular day-to-day contact.

At the very beginning of DESY, Peter Stähelin and Martin Teucher, who had come with Jentschke from the USA, were essential. DESY has always had a cohort of leading scientists who are essential for the long-term success of the laboratory. Most of them served for a number of years as members of the directorate. Among them are the physicists Erich Lohrmann, Gustav Weber, Günther Wolf, Paul Söding, Johann Bienlein and Gerhard Horlitz.

From the technical side there is of course Gus Voss, as well as Donatus Degele and Hermann Kumpfert. The administration is often overlooked and always essential, and one person who defined it for many years was Heinz Berghaus followed by Senatsdirektor Richard Laude and later by Wolfram Schött who came from the research ministry at Bonn and we became friends. In various functions supporting the directorate were Wolfgang Grillo, Peter von Handel, Helmut Krech, and Gerhard Söhngen.

When you hold a prominent position, you need somebody to protect you from too many meeting requests, someone you can trust implicitly to represent you to see you: a *Chef de cabinet* as it is called today, but at that time the person who carried out this task was my secretary, Karin Schmöger who made my life bearable. Last but not least, I would like to acknowledge those who became my successors as chairs of the DESY directorate Volker Soergel, Bjørn Wiik, Albrecht Wagner and Helmut Dosch, the latter two of whom gave DESY a new direction in research.

During the construction of PETRA there were many discussions as to whether DESY should legally become an international laboratory. This did not happen, and DESY remained, at least officially, a national lab. However, the international participation in PETRA experiments was broad, and many scientists from all over the world contributed. An outstanding one is Sam Ting."

References for Gluon Discovery

1. Schopper H (1980) Two years of PETRA operation, DESY GD 80/02, Dec 1980
2. Soding P (2010) On the discovery of the gluon. Eur Phys J H 35(1):3–28. https://doi.org/10.1140/epjh/e2010-00002-5
3. Wu SL, Zobernig G (1979) A method of three-jet analysis in e + e annihilation. Z Phys C Part Fields 2(2):107–110. https://doi.org/10.1007/bf01474124

Chapter 7
Director-General of CERN

On 1 January 1981, Herwig Schopper became Director-General of CERN. It was a role that his entire career had prepared him for. Years of experience in managing and establishing laboratories across Germany, culminating with his time as chair of the DESY directorate gave him the managerial experience he needed, while his sojourns in Stockholm, Cambridge and Cornell had kept his research instincts sharp. The final ingredient in his preparation for what is arguably the biggest job in particle physics is the time he'd spent at CERN in the 1960s and 1970s learning about this unique institution and its equally unique culture. "Before being able to take full responsibility for an organisation as diverse and complicated as CERN," he explained, "one has to get to know it from the inside, as well as understanding the world outside. When I came to CERN in 1970 as leader of the Nuclear Physics Division, I took over from the Swiss physicist Peter Preiswerk. It was a big honour to become his successor, and I learned a lot about CERN over those three years."

Herwig headed the Nuclear Physics Division for two years before taking on the role of experimental physics coordinator for one more year before he moved to DESY. "Bernard Gregory was the Director-General at that time," recalled Herwig, "and my immediate boss was the head of what was known as the Physics I Department. Giuseppe Cocconi had just handed over to Jack Steinberger. Both were outstanding physicists, Steinberger later received the Nobel Prize, and both were remarkably strong personalities. Neither liked to do administrative work, which resulted in me taking on some of the tasks of the Research Director."

Herwig remembers one young physicist from his time as division leader particularly strongly. "One day a relatively young guy, but very ambitious, came to me and asked for money," recalled Herwig. "His name was Carlo Rubbia, and when I asked what experiment he was involved in, he listed about a dozen. He was always full of new ideas, and when I asked why he wanted to start yet another experiment, we had a rather heated discussion. Carlo had a fiery temperament, but in the end we always reached an agreement, and Carlo later succeeded me as Director-General."

© The Author(s) 2024
H. Schopper and J. Gillies, *Herwig Schopper*, Springer Biographies,
https://doi.org/10.1007/978-3-031-51042-7_7

Fig. 7.1 Giuseppe Cocconi preparing for an interdivisional meeting at CERN in 1975. "Cocconi had a great sense of humour," recalled Herwig. "He would often say to me: "the world is mad except you and me, and I'm not sure about you"" (©CERN, All rights reserved)

When Herwig arrived at CERN in 1970, there was no collider in operation at the laboratory: the physics programme was built around fixed-target experiments. The Intersecting Storage Rings (ISR) were under construction, and they would bring collider physics to CERN for the first time in 1971, changing the laboratory, and the field, forever. In fixed-target experiments, the particles that emerged were analysed using detectors built around large and complicated magnets. This led to another memorable interaction for Herwig. "It turned out that no new experiments could be approved because there was a limited number of analysing magnets available," he recalled. "So one day I went to see Bernard Gregory, bypassing the directors, and I asked if we could buy a few more analysing magnets so we could approve more experiments. Gregory laughed and said, 'Look, somewhere there must be a limitation. If it's not the analysing magnets in the end it might be the capacity of the cafeteria or the number of parking spaces.' Many years later, I thought about that remark very, very often because it taught me that sometimes you have to make decisions against the obvious rational arguments. Gregory was a wise man with great foresight. Ironically if you come to CERN today you might easily get the impression

Fig. 7.2 A formidable gathering in the CERN auditorium. From second left to right, James Cronin, Jack Steinberger and Leon Lederman—Nobel Prize winners all. Cronin is in conversation with CERN's Jean-Pierre Stroot (©CERN, All rights reserved)

that cafeteria and parking capacity are indeed the present limits of CERN." In the end, no more analysing magnets were bought in the early 1970s, and Herwig does not recall whether Rubbia got his extra experiment. The arrival of colliders, however, would soon mark a new departure for both men.

During his three years as a division leader at CERN, Herwig learned to admire the vision of the laboratory's founding fathers. He came to appreciate that the procedures they had put in place provided a powerful blueprint for international collaboration in science and were essential for the success of the organisation. "One was the rule that there should be no national quotas, either for the employment of staff or for the adjudication of contracts to companies," he explained. "The guiding principle should be that the best scientists are welcome if they can contribute to the scientific success of CERN." This principle applied even beyond the boundaries of CERN's member states. "The participation of Polish and Israeli scientists, for example, was possible right from the beginning and they made essential contributions," he continued. "In the meantime, these countries have become member states." Herwig also came to appreciate CERN's approach of placing orders with the companies best able to fulfil the contract, rather than applying some form of *juste retour*. "Contracts were given to the lowest bidding firm that was technically able to fulfil the contract," explained Herwig. "There is no rule of *juste retour* at CERN as there is in some other international organisations—that only increases the cost since a firm might know that it is their turn for the next contract."

When Herwig left CERN for DESY in 1973, his apprenticeship for the top jobs in the field was complete. He maintained close relations with CERN, being elected to the organisation's Scientific Policy Committee in 1978 and chairing the ISR Committee from 1973 to 1976. "This was an extremely interesting time from the physics point of view," he said, "and it also provided me with the opportunity to get to know many people who were essential during my time as DG."

Electing the New Director-General and Reunifying the Lab

The circumstances surrounding the election of Herwig as Director-General were somewhat unique in the history of CERN. In 1971, with the approval of the SPS, CERN had been split into two laboratories, CERN I and CERN II, with two Directors-General (see Chap. 4). A few years later, however, the CERN Council had taken the decision to reunify the two into a single administrative structure. "Although CERN had officially been recombined in 1976 with the appointment of John Adams as Executive Director-General and Leon van Hove as Research Director-General," explained Herwig,"there still remained much to do to unify the two labs when I arrived in 1981." The new Director-General would also be tasked with developing a full proposal for a new project to succeed the SPS, and then building it assuming that it was approved.

The new Director-General would also assume office in the first year that the SPS was to run as a collider, and to complicate matters further, the Council had decided that CERN's budget would henceforth remain constant in real terms. Finding a single candidate with the experience to manage such a complex agenda was a tall order, and in the end, it was Herwig's combination of scientific track record coupled with managerial experience, both inside and outside CERN, that won the day. He took up the post on 1 January 1981.

Although the SPS had been sited adjacent to the existing CERN site, by the time that decision was taken the CERN Council had already appointed John Adams to be Director-General of the new laboratory. From 1971, therefore, CERN was run as two separate entities, the original Lab I, which was responsible for the basic programme under Willibald Jentschke, and Lab II, which had the task of building the SPS under Adams's direction. When Jentschke's term ended, the labs were partially reunited with Adams as Executive Director-General, and Leon van Hove as Research

Director-General. It was when their mandate came to an end that the CERN Council had decided to reunite the laboratory fully under a single Director-General. This became one of Herwig's most difficult tasks.

The usual procedure for selecting a Director-General of CERN involves closed-door discussions throughout the fourth year of the current mandate. Candidates are put forward and debated, each member state delegation advances arguments, consults with ministries and gradually the process converges on a single candidate who is appointed by a unanimous vote of the council at its December meeting. In 1979, Herwig Schopper was a clear front runner. It was already evident by this time that the SPS's successor would be an electron machine, a large electron–positron collider (LEP), and as head of a laboratory that had made its reputation on electron machines, he was a natural choice to lead CERN from its proton-dominated past to a future electron–positron facility.

All was not done and dusted, however, since despite Herwig's clear suitability for the job, unanimity was proving elusive: Italy was not convinced. The Italian delegation was worried that Herwig would not do everything possible to get the new project to Geneva, but rather that he would favour DESY. The Italians also had their own candidate, Antonino Zichichi. Things seemed to be at an impasse, and the clock was steadily ticking towards 1 January 1981, the start date for the new mandate. Things were resolved by a caller at Herwig's door.

"One Saturday afternoon in December 1979, the doorbell rang," recalled Herwig, "it was Nino Zichichi. We were old friends, and he'd come to discuss the Italians' concern that my heart would still be beating for DESY, and that I would not fight for LEP in Geneva. I told Nino that LEP was too big a project for DESY and that I would fight as hard as I could to get the project to Geneva if I were made DG." The next time that Herwig visited CERN for a committee meeting, Zichichi organised a meeting for him in a hotel with the Italian delegate to the CERN council, Umberto Vattani, and it was agreed that Herwig should be invited to the next meeting of the Committee of Council on 29 February 1980. After his presentation, and a quick phone call to Rome, the vote was held and a press release was issued noting that: "the 12 delegations unanimously supported the appointment for five years of Professor Herwig Schopper to the post of Director-General of the organization from 1st January 1981."

"The whole operation was guided very skilfully by the Council president, Jean Teillac, who even had the courage to send me a letter of appointment dated 29 February 1980, which made it possible to look immediately for my successor at DESY," explained Herwig. "My nomination was formally approved in a special Council meeting on 25 April, on which occasion the Italian ambassador expressed warm congratulations." Herwig acceded to the top job in particle physics in Europe, with a metaphorical mountain to climb to get LEP approved and constructed during his five-year term. He spent the rest of 1980 preparing to hand over the reins at DESY to Volker Soergel, and to take over the running of CERN from 1 January 1981.

He also immediately started work with Adams and van Hove on a new proposal for LEP to put to the Council in June 1980 as the delegations had requested. "I developed a very good working relationship with Adams and van Hove," he recalled.

"We prepared a new proposal based on an earlier study known as the pink book, but with an important difference. Instead of building a whole new pre-accelerator system, CERN's existing accelerators, the PS and SPS, would be used as pre-accelerators." This brought the cost of the project down to from well over a billion Swiss Francs to 950 million. It was an important step, but LEP's approval would still require a lot of hard work once Herwig took up office at CERN. In 1980, however, developing the proposal along with preparations for the transition at DESY kept him busy. "Sometimes I had to do the work of two DGs and even start the new job almost a year early without being paid," he said with a smile.

The LEP Proposal

In March 1975, particle physicists from around the world gathered in New Orleans for a topical seminar to discuss the future of the field. Herwig was there in his capacity as chair of the DESY directorate. "The main outcome of the meeting was that the next big facility should be a very big accelerator (VBA), constructed as a world machine," he recalled, "and that an international study group should be established with its headquarters at CERN." Two months later, the subject was being debated in the Scientific Policy Committee at CERN, which delivered a report to the laboratory's Committee of Council, and before long, the VBA had morphed into the Large Electron–Positron collider, LEP. "It's not clear who at CERN first suggested the construction of such a machine," said Herwig, "but already in New Orleans, the talk was of an electron machine, and from a European perspective, it seemed like a natural continuation of the tradition that began with AdA and ADONE in Italy, ACO in France and DORIS and PETRA at DESY. Anyway, CERN took the initiative before the international study group that the New Orleans meeting had advocated really got off the ground."

Studies for LEP got underway in earnest at CERN in 1976, but the idea was not universally popular at a laboratory that had made its name with proton machines. The SC, PS, SPS and ISR were all proton machines, and the SPS was soon to be converted into a hadron collider, colliding protons and antiprotons in a bid to discover the W and Z particles, carriers of the weak nuclear interaction. The SPS collider project was approved by the CERN Council in 1978, and it carried with it the hopes of a generation of experimental particle physicists. There was good reason for optimism: in 1973, a team led by Frenchman André Lagarrigue had identified so-called weak neutral currents for the first time. Although not a direct observation of the neutral carriers of the weak force, the Z particles, this was compelling evidence for the electroweak theory advanced by Glashow, Weinberg and Salam, and discovery was only a matter of time and energy.

Perhaps paradoxically, the success of CERN's hadron machines in advancing the emerging standard model of particle physics, in tandem with theoretical developments, pointed inexorably towards an electron machine as the next facility that the world's particle physicists would need. While the SPS collider was widely expected to discover the W and Z particles, it would not produce these particles in sufficient numbers to study them in fine detail, and nor would it provide the pinpoint precision that would be required. Deliberations on a possible proton–proton or electron–proton collider at CERN soon petered out faced with the undeniable logic of LEP. A series of designs for LEP was advanced, mainly under the guidance of John Adams, and each was dissected in CERN's Scientific Policy Committee (SPC), of which Herwig was a member, until a feasible, and affordable, proposal emerged.

The first decision the accelerator designers had to make was the energy reach of the new machine. It would have to be able to produce Z particles in the first instance, and although they had yet to be discovered, their mass was expected to be below 100 GeV, pointing to a beam energy of 50 GeV per beam. To study the electrically charged W particles, however, which would be produced in pairs at an electron–positron machine, the required energy would have to be roughly twice as much. Then there was the question of whether to build a linear collider or a circular one. Here, the economic argument prevailed: for energies below around 300 GeV, a circular machine was much more cost-effective, and would have the advantage of being able to collide the same beams many times over. "It is not surprising that it took some time to agree on a final energy for LEP," said Herwig. "The energies suggested fluctuated up and down, leading to different circumferences for the ring being advanced." The challenge was exacerbated by the limited space available between Lake Geneva and the Jura mountains, with the unknown geological challenges their limestone structure presumably concealed.

The first study group convened in 1976 opted for a design with a 50 km circumference and a beam energy of 100 GeV, sufficient to produce W particle pairs, but it had significant drawbacks. When the LEP 100 proposal went to the SPC, it was quickly rebuffed. Although able to do everything the physicists needed, it would have involved tunnelling deep under the Jura mountains with long access tunnels and the cost was simply too high. The LEP study group was sent back to the drawing board.

In 1978, the group presented a proposal for LEP 70, which would come to be known as the blue book study. It proposed a ring of 22 km, which would avoid the worst of the difficult geology under the Jura, and accelerate beams to 70 GeV, with a later upgrade to 100 GeV per beam when superconducting accelerating cavities became available. It too was rejected on the basis that higher energy with conventional cavities was desirable, and that the experimental areas foreseen were not large enough. A pink book succeeded the blue, this time with Herwig as a co-author, as by this time he'd been elected to be the next Director-General of CERN, and much of 1979 was spent discussing its feasibility. The pink book design upped the circumference from 22 km to 30.6 km, and the beam energy with conventional accelerating cavities to 86 GeV, but it was back to unknown geological conditions and potentially too high a cost. It was not until Herwig was firmly settled in office that the final design, with a 27-km ring, would be approved, allowing construction to get underway.

A New Style for CERN

Before LEP, every new CERN project had been funded independently of CERN's basic research programme, with CERN member states in principle being able to opt in or out as they saw fit. With LEP, however, that was about to change. Long discussions with member state governments made it clear that some countries wanted to participate in the basic programme but not in the LEP project, while at the same time, the LEP proposal relied on using existing CERN facilities that were part of the basic programme. "An impossible situation seemed to be emerging," said Herwig, "so I said to the Council that LEP should become part of the basic programme with the consequence that it would have to be approved and funded by all the member states." As a result, the December 1980 Council meeting concluded with the statement that: "If the inclusion of the LEP project phase 1 in the scientific activities and long-term budget estimates is agreed by the Council with no member state voting against, this will constitute approval of the LEP project phase 1." This made Herwig's task clear: he needed to obtain a unanimous vote in the Council, with possible abstentions, for the LEP project to go ahead. Furthermore, he was asked to present a definite proposal to the Council for its June 1981 meeting, along with a financing scheme that would integrate LEP into the CERN basic programme with no additional funds. The Council was making it clear that the days of extra funding for new projects was over.

Hardly had Herwig got his feet under the desk of his 5th floor office in CERN's main building than he was off on the road touring the capitals of CERN's member states to drum up support for the project. "Most of the opposition came from colleagues in other fields," he recalled. "They feared that it would not be possible to build LEP within a constant budget, and that their budgets would suffer as a consequence." To win them over, Herwig realised that he had to change the mind-set at CERN. "CERN was organised into divisions, each one responsible for accelerators, or physics or services," he explained. "Each was highly competent in its domain, but they all worked quite independently of each other. Their budgets were known only to the DG and a few other people. I realised that for LEP to be built under the conditions required by the Council, I'd need to organise it as a project that cut across all the divisions." The challenge was exacerbated by the fact that Herwig also had two independent laboratories to unify into one. "John Adams was a very strong personality," he explained, "and he commanded great loyalty among his staff."

The accelerator builders at the former CERN II expected an accelerator physicist to be appointed to lead the LEP project, but Herwig surprised the community by appointing an experimental physicist, Emilio Picasso, instead. "Many were disappointed," he recalled, "but Emilio had worked on experiments that required close collaboration with the technical divisions, and he enjoyed enormous respect not only among the physicists, but also among the accelerator engineers." In addition to appointing Picasso as project leader and a member of the directorate, Herwig also established a new division for LEP, and appointed the leading accelerator physicist, Günther Plass, to be both the division's leader and Picasso's deputy. It was an arrangement that worked well. "Emilio Picasso was the right person to bridge the gap between the staff of CERN I and CERN II. He did a great job, and we became close friends," said Herwig. "This proved important when we had difficult decisions to take together as the project advanced."

Herwig's mandate was a pivotal one for CERN, marking the transition from a constantly growing organisation to a steady state. Up to the 1980s, new projects bene-fitted from extra resources, and once a project was approved, CERN was responsible for providing all the necessary infrastructure, for both accelerators and detectors. The laboratory's personnel and budget had grown steadily since the organisation's creation in the 1950s, and in addition there were projects, such as the Big European Bubble Chamber (BEBC), that were funded outside the core budget by groups of member states with a specific interest. In the case of BEBC, these were largely France and Germany. When Herwig presented the project for LEP that he'd developed with

Fig. 7.3 Herwig in conversation with Emilio Picasso at the XI International Conference on High Energy Accelerators, which was hosted at CERN in July 1980 (©CERN, All rights reserved)

Adams and van Hove to the Council in June 1980, the Council made it very clear that this situation had to change. They wanted LEP built within a constant budget, and with no extra personnel. Furthermore, they wanted that budget to be lower than the one he had inherited: 629 million Swiss Francs in 1981.

At 1300 million Swiss Francs, the budget for LEP in the pink book design was clearly impossible under such conditions, and a more modest proposal needed to be developed. Consultation with tunnelling experts led to the conclusion that the tunnel needed to be pulled back from under the Jura mountains to avoid financial risk. "After long and sometimes heated discussions, we decided to reduce the circumference of LEP to about 27-km, and move it somewhat out of the Jura," Herwig explained. "Fortunately, the clever LEP design group found a way to keep the phase 1 beam energy of 50 GeV, even with the reduced size." Not everyone was convinced that Herwig had gone far enough, and he received letters from people including John Adams and Carlo Rubbia urging him to move the machine completely from under the Jura by reducing the circumference to around 20 km, but Herwig stuck to his guns. "These opinions gave me some headaches and sleepless nights," he recalled, "but finally I decided to maintain the tunnel circumference at 27-km." Herwig was thinking ahead. "Reducing the size would not have impaired LEP's performance considerably, but I decided to take the risk attached to tunnelling under the Jura because one thing we could not change in the future is the size of the tunnel," he explained. "I thought we should have the biggest tunnel possible, and that proved to be an important factor when planning the LHC project, making it possible to reach interesting energies."

Herwig approached the first big test of his mandate, the June 1981 CERN Council meeting, armed with a new design for LEP, this time contained within the pages of a green book, and ready for some tough discussions. The green book design, which came to be known as the stripped-down LEP, presented the project as an evolving machine to be completed in two phases, with phase 1 being a factory for Z particles. Phase 2, which would bring superconducting radiofrequency (RF) technology to bear and produce W particle pairs, was left for a later decision. Herwig planned to draw on the expertise developed in part by his teams when he was in Karlsruhe to develop that technology while phase 1 was underway.

Herwig had accepted the inevitability of the Council's desire to construct the machine under a constant budget, and against the advice of his colleagues, went to the Council meeting prepared to accept stable funding at 629 million Swiss Francs per year, which he considered to be realistic if the timescale for construction was extended by one year. If that was not enough, time would provide the contingency, because Herwig had already accepted that further funding would not be forthcoming. He also recognised that even if extra funding were possible, it would not be in CERN's interests because it would pit particle physics against other fields of science, and that would be detrimental to the scientific endeavour as a whole.

As well as reducing the circumference of the machine to 27-km, further savings on civil engineering were made by foreseeing four experimental halls instead of the eight in the pink book, and, in a new departure for CERN, only a token amount was included for experimental infrastructure. The cost of the experiments would have to be borne by the collaborating institutes that proposed them. This was a completely new approach for CERN.

Herwig emerged from the Council meeting bruised but not beaten. He had achieved neither approval for LEP, nor the constant budget that he'd argued for, "Some delegations wanted to come down to 610 million Swiss Francs per year," he recalled, "and we ended up with 617. That might not seem like a huge difference compared to what I asked for, but over the eight-year construction period of LEP, it would add up to 96 million." Even Herwig's attempts to have the materials part of the budget indexed were met with a resolute "no" from the council, he'd have to go back each year to argue for that. The bid to get LEP approved fared only slightly better. Eight of CERN's 12 member states voted in favour of the stripped-down LEP, with Denmark giving a positive vote ad referendum. "It was not completely clear what that meant," said Herwig. Three-member states asked for more time to reach a decision. "For a few months, the trembling continued, and I spent much time visiting member state capitals," said Herwig, "The big sigh of relief only came at a special Council meeting on 30 October when LEP was approved unanimously."

Herwig had succeeded, but at a price. These Council meetings had changed the way CERN operated for ever. No longer would CERN enjoy rising budgets and personnel, and no longer would CERN cover the cost of the experiments it hosted. Although indexation has been granted since Herwig's time in office, in real terms, CERN's budget has not risen since the 1980s, and the personnel has fallen while the user community has grown dramatically. Herwig was not a popular Director-General at this point of his mandate, but time has vindicated his approach. "I had no choice but to accept what the Council wanted," he said, "but it went against all the tradition of CERN." There were vociferous calls from the CERN staff for his resignation. "But I told them that I could only step down once, and that if I did, the situation would not change," he recalled. An uneasy truce was declared as CERN's new era began.

The price to be paid was a significant one, with many parts of the CERN programme being cut back or stopped altogether. "The ISR, the only proton–proton collider in the world, had to be stopped in 1983, a particularly painful decision," recalled Herwig. Along with the ISR went BEBC, much of the fixed target programme at the PS, and a significant amount at the SPS. The operation of the synchrocyclotron, CERN's first accelerator was reduced from 6000 hours per year to 4000, and even accelerator R&D, essential for the future of the laboratory, was cut back to focus exclusively on the superconducting RF technology that would be essential for LEP's phase 2. A mere 1% of the CERN budget went into the R&D needed for the long term. "I lost many friends," said Herwig ruefully, "but fortunately I later regained most of them."

Fig. 7.4 Herwig in discussion with representatives of the CERN Staff Association early in his mandate as Director-General (©CERN, All rights reserved)

A Nobel Discovery

One significant survivor of the cuts was the SPS collider programme—it had been that or the ISR, and the higher-energy machine was the obvious choice. Herwig was nevertheless in for a surprise. The project to convert CERN's big machine into a collider had been taken before the changes ushered in in 1981, and as Herwig was to discover, it had had its own funding challenges. "When I became Director-General, I learned that this special programme had never been formally approved at the Council," he recalled. The conversion of the SPS to a collider was carried out within CERN's accelerator development programme, and the Council had not formally allocated resources for it. A similar story was true for the experiments: two underground area (UA) experiments had been discussed in the scientific committees:

UA1, under the direction of Carlo Rubbia, and UA2 with Pierre Darriulat as its spokesperson, but neither was fully funded. "When I came in Rubbia and Darriulat asked for more money to complete the equipment of these two experiments," Herwig recalled. "There was no money for them in the budget, but certainly it was clear that these were key experiments, so we had to go on." Herwig managed to eke out some funding for UA1 and UA2, with the Council's blessing, but much of the funding for these flagship experiments was found outside the CERN budget. "This is very interesting because UA1 and UA2 are the first examples of two big collaborations at CERN that were partially financed through outside contributions," he explained. "They became the prototypes of the big experiments at LEP and the LHC. Although they were formally under the direction of CERN, under the strong leadership of Rubbia and Darriulat, they achieved a certain independence."

Data-taking at UA1 and UA2 began in 1981, and the following year the two experiments had clear signs of new particles. "Just before Christmas, Carlo Rubbia and Pierre Darriulat came to see me in my office, independently, and each showed me their data very confidentially," said Herwig. "It was clear they both had evidence for the W."

With the exception of the prime minister of one CERN member state, who had requested to be informed directly (see Section "A Prime Ministerial Visit", In his own words: A prime ministerial visit) the world discovered the news through a CERN press release issued on 21 January 1983. Discovery of the Z boson followed in the summer. The following year, the discoveries of the carriers of the weak interaction were crowned with the award of the Nobel Prize to Carlo Rubbia and Simon van der Meer, the engineer whose invention of a technique known as stochastic cooling had made the endeavour possible. Stochastic cooling enabled beams to be stored in such a way as to increase the number of collisions dramatically.

It is rare for a Nobel Prize in Physics to be awarded so soon after the research is conducted, and, with the prize going to a maximum of three individuals, increasingly rare for experimental physicists to be rewarded. UA1 and UA2 were both collaborations numbering dozens of researchers, but with a little help from Herwig, the Nobel committee found good reason to single out Rubbia and van der Meer.

"A delegation from the Nobel committee came and asked to see me," he explained, "we had a nice lunch and they told me privately and in strict confidence that there was a proposal to give the Nobel Prize to Carlo Rubbia and Simon van der Meer, but the committee had problems agreeing to it." First, they discussed van der Meer. Famed for his tendency to avoid the usual publication route, van de Meer's work on stochastic cooling had been published as internal CERN reports, and not in peer reviewed journals. This gave the Nobel committee a problem, as the Nobel recognises only work that has been subject to peer review. That was an easy problem for Herwig to solve. "The next day I called Simon to my office and told him to write a summary report about his beam cooling method and get it published in a peer reviewed journal," said Herwig. "Of course, there was no problem passing peer review, and he did it within two weeks."

Carlo Rubbia's case proved more of a challenge and raised an issue that still bothers Herwig today. "They told me that the Nobel Prize can be given only for

Fig. 7.5 Herwig presides at the announcement of the discovery of the W boson in January 1983. He is flanked on his right by Carlo Rubbia and Simon van der Meer, and on his left by Erwin Gabathuler and Pierre Darriulat (©CERN, All rights reserved)

a discovery or invention, and the W and Z particles were neither, because they had been predicted by theory," explained Herwig. "I suggested that they could give him the prize for his whole career, but they ruled that out because according to Alfred Nobel's testament, the prize is always given for something specific, so then I suggested awarding it to the whole collaboration, but they turned that down as well." Herwig argued that many theoretical predictions turn out to be wrong, and without experimental verification, they are worth very little, but the Nobel committee members told him that the prize can only be given to a maximum of three people, according to the will of Nobel himself. "I told them that there's nothing in Nobel's testament that specifies three people only," said Herwig, "and after long discussions they told me the real reason—they wanted to keep the Nobel Prize popular." The delegation from Stockholm explained to Herwig that as the subject matter of the scientific Nobel Prizes was frequently beyond the grasp of the general public, they wanted to create heroes. "This is a problem that is still not solved," said Herwig, "the same difficulty arose with the Higgs particle, when the question came up again, and again they gave the prize to the theorists only and not to the experiments, but I repeat the theories would be worth nothing if they had not been verified by experiments. So, I regret that the big collaborations of today, starting with the collaborations at DESY discovering the gluon, then UA1 and UA2, going on to LEP and now the LHC collaborations cannot receive the Nobel Prize. It's a pity because progress in science is possibly only through strong collaboration between theory and experiment. Both are necessary and the public should be properly informed to understand it. What can we do?"

Herwig had to think hard to come up with a way of convincing the committee members that they could award the prize to Rubbia. He hit on the idea of using that same justification as that used for van der Meer. "I told them that if they were to give the prize to Simon van der Meer for cooling, they could use the same argument for Carlo Rubbia," he recalled. Cooling was the key to the SPS collider's success, it allowed sufficiently intense beams of antiprotons to be accumulated to generate enough collisions to make the discovery possible, and although it was van der Meer's technique that was used, Rubbia had proposed one of his own when he lobbied to convert the SPS into a collider. "Carlo Rubbia had proposed a different method," Herwig explained, "he proposed injecting heavy lambda particles into the SPS, which would decay into protons and antiprotons. This idea was never realised because it turned out that Simon van der Meer's method was much more efficient, but it allowed me to argue that they could give the prize to Rubbia as well. I also argued that they should at least mention the UA1 and UA2 collaborations."

How much influence that visit to Geneva had on the Nobel committee's deliberations will remain forever sealed behind closed doors in Stockholm, but the prize did indeed go to Rubbia and van der Meer in 1984, with a citation that read: "This year's Nobel Prize for Physics has been awarded to Professor Carlo Rubbia and Dr Simon van der Meer. According to the decision of the Royal Swedish Academy of Sciences the prize is given "for their decisive contributions to the large project, which led to the discovery of the field particles W and Z, communicators of the weak interaction." The presentation speech delivered by Professor Gösta Ekspong of the Royal Swedish Academy of Sciences also mentioned the two collaborations, albeit not by name. "Perhaps there is good reason for awarding the scientific Nobel Prizes to individuals instead of collaborations, unlike the Peace Prize, which is awarded in Oslo and can go to organisations. Individual winners do achieve a certain degree of recognition, which they can use to lobby governments to support science," concluded Herwig, "but there should be some way to recognise the work of the large collaborations as well."

Spain Re-joins CERN: An Excursion in a Wheelchair

The advent of the LEP project brought growing prestige to CERN as a symbol of European success, triggering an enlargement of the laboratory's membership that continues to this day. The first new country to join, or more accurately re-join, was Spain. A member state of CERN from 1961 to 1968, by the 1980s, Spain was ready to come back to the fold. "A key person in the renaissance of particle physics in Spain was Juan Antonio Rubio who had studied at the university of Madrid and joined experiments at CERN," said Herwig. "Thanks to his influence the Spanish central government took a positive attitude. Parliamentary approval nevertheless turned out to be rather complicated because Spain is a collection of 17 autonomous regions, and they all had to agree."

Early in the new year of 1982 the Schopper family went on a skiing holiday near Chamonix in the French Alps. "I made the big mistake of trying to compete with my children," recalled Herwig. "I had taught them to ski but by now they were better than me. As a result, I had a bad accident and broke my leg in a very complicated way." Herwig was taken to the Cantonal Hospital in Geneva, where he found himself lying in a hospital bed parked in a corridor, near the bottom of the priority list. "It was the middle of the ski season, I think the car show was on in Geneva, leading, ironically, to many traffic accidents and so the hospital was very busy and the wards were full," said Herwig. "One day my wife came to look for me and she couldn't find me. Since mobile phones had not yet been invented, it was difficult to stay in contact with the outside world." By this time, Herwig's daughter Doris, a doctor who had very good relations with the head of the department for internal medicine, was away on a mission for Médecins Sans Frontiers in Central America. "So she couldn't help me," explained Herwig. "On the third day of waiting, though, one of the student doctors asked me whether I was Doris's father, so at least I had somebody to talk to. He even talked to the head of his department to get me a bed, which was quite revolutionary since patients were never housed in departments other than the one that was responsible for them." Eventually, Herwig's leg was operated on, and he left the hospital in a wheelchair with his leg in plaster, ready to negotiate Spanish membership of CERN.

While still in hospital, Herwig received a message that the Spanish government desperately wanted him to visit Madrid to clarify some unsettled issues concerning Spain's membership of CERN. It seemed that there was a one-week window in which to secure the approval of parliament, and if that window was missed, the opportunity might well disappear. "So, despite the fact that none of my family could accompany me I agreed to go. In the wheelchair, with my leg in plaster," said Herwig. "It turned out that the questions were very easy to answer, and the negotiations with the minister were settled in just a few hours."

Herwig's reward was not just that Spain formally re-joined CERN on 1 January 1983. "The minister asked whether I had been to Toledo," explained Herwig. "Since my answer was negative, he had me taken there by car, where I found out that museum visits are much less tiring in a wheelchair. They also put me up in one of the famous

Paradores, which had no lift. Getting up to the first floor on my broken leg was a nightmare. It was a visit I'll never forget."

Herwig stayed in close contact with Juan Antonio Rubio, who became a division Leader at CERN and went on to be employed by the Spanish government in various capacities before being appointed Director-General of the Centre for Energy, Environmental and Technological Research (CIEMAT) by the Spanish Ministry of Education and Science in 2004. "To my great regret he passed away much too early in 2010," said Herwig, "and I lost another friend."

Portugal: Much Work in a Beautiful Country

Three years to the day after Spain re-joined CERN, Portugal also became a CERN member state. Once again, the driving force was an individual physicist with a big vision for science in his country.

CERN sets conditions for membership in order to ensure that prospective new member states are able to benefit fully from their investment in the laboratory. One of these is that the country must have an active particle physics research community. "Universities in Lisbon and Coimbra had a famous history in physics but the whole structure was old-fashioned and not adapted to modern science," recalled Herwig. "A dynamic physicist, Mariano Gago, who had spent some time working in France wanted to change this." An electrical engineer by training, Gago had turned his hand to particle physics at Paris's prestigious Ecole Polytechnique. "It was largely thanks to Mariano's efforts that Portugal became a member of CERN," said Herwig, "along with Gaspar Barreira and Armando Policarpo he also established LIP, the Laboratory of Instrumentation and Experimental Particle Physics, as a focal point for Portugal's involvement in CERN experiments. He became a great personal friend."

Gago went on to become Portugal's Minister for Science, Technology and Higher Education, and was succeeded at LIP by Gaspar Barreira. This allowed him to further his advocacy for science. "Mariano was instrumental in transforming Portugal into a modern country as far as science and education are concerned," said Herwig, "and I remember him being a very active member of the CERN Council during my mandate as Director-General."

After Herwig's retirement, Gago joined the long list of people seeking to tap into his long experience. "He asked me to join an advisory committee for LIP," Herwig recalled. "I accepted and that led to a ten-year association with Portuguese science." Herwig went on to chair the LIP committee, and later joined a government committee set up to advise on the development of science in Portugal. Herwig was active in the subcommittee on physics, which he also went on to chair. "I had the opportunity, and the responsibility, to influence activities in the most prominent Portuguese universities," he recalled, "which meant hard work and difficult discussions, but it was worth it. My contacts with Portuguese colleagues were always based on mutual respect and recognition, and never derailed into fruitless argument."

Fig. 7.6 A visit from the Portuguese Minister for Foreign Affairs, Jaime Gama in 1985. Jose Mariano Gago is standing centre, behind the minister (©CERN, All rights reserved)

Mariano Gago passed away in 2015 leaving a strong scientific legacy in Portugal and beyond. Today, LIP is an institution with a large and diverse programme, participating in several international collaborations and with activities ranging from IT to medical applications. "When Mariano's health no longer allowed him to fulfil all his obligations, he was greatly supported by his colleague Pedro Abreu," recalled Herwig. "This was no accident, Pedro is outreach coordinator at LIP, and Mariano was a great supporter of education and outreach. I owe Mariano special gratitude for introducing me to Portuguese history. He explained to me the importance of Henry the Navigator who, it is said, was a strong supporter of science and may even have set up a scientific school for navigators. Mariano used to joke that Henry was centuries ahead of CERN! With Mariano, I lost a real friend and I much regret that he could not enjoy all the successes that he had initiated and completed for longer."

Fig. 7.7 Italy's Foreign Minister, Giulio Andreotti (centre) also visited in 1985, accompanied by Antonino Zichichi (left) (©CERN, All rights reserved)

Water at the End of the Tunnel

When civil engineering for LEP got underway in 1983, it was the largest construction project in Europe until work began on the Channel tunnel in 1988. Hundreds of workers descended on the Geneva region to work on the project, and temporary accommodation sprang up to house them all in the largely rural Pays de Gex region of France. Before that could happen, however, there were some more tweaks to be made concerning the machine, above all its precise location. The tunnel's placement was once again shifted to move more of it from under the Jura, leaving less to be excavated in the difficult terrain of the mountains, and reducing the maximum depth of the access shafts from 600 metres to 150.

The trigger for the change came in the form of a note in Italian from Emilio Picasso to Herwig. "Italian was the language he always used when he was very worried," said

Fig. 7.8 Herwig (right) discussing the LEP project with the President of France, François Mitterand, and the President of Switzerland, Pierre Aubert, on the occasion of the LEP ground-breaking ceremony on 13 September 1983 (©CERN, All rights reserved)

Herwig, who as a consequence took the note very seriously. In it, Picasso summarised the geological risks identified by an expert from the prestigious Swiss Federal Poly-technic Institute (ETH), in Zürich, and it made for sobering reading. Water up to 20 atmospheres in pressure had been identified through test borings, along with several geological faults that the green book design would have to cross. The experts from ETH had identified a series of factors that could lead to a delay of up to 16 months at a cost of up to 17 million Swiss Francs. This is not a risk that Herwig was prepared to take. "We decided to displace the main ring towards the east by several hundred metres," he explained, "bringing it more out of the Jura and closer to the airport." The resulting position left only three kilometres of the tunnel under the Jura to be exca-vated by explosives, with the rest in molasse sandstone, which could be excavated using a tunnel boring machine.

There was another advantage to the new position too. "The land needed for access shafts had to be acquired by the two host states," explained Herwig. "In Switzerland, a popular vote is necessary for the acquisition of land by government, and that could have taken a long time. The new position of the ring had just one access point in Switzerland, and that was on land that had already been made available to CERN for the SPS. All the other access points were situated in France, with the shafts at points six and eight just a few metres outside Switzerland." The precise position of the ring was kept under wraps for a long time, and was known only to Herwig, Emilio Picasso and a few other experts at CERN because they thought it could have

Fig. 7.9 In conversation with German delegate Josef Rembser in the margins of a CERN Council meeting in 1983. In the following years, Rembser went on to be vice-president and then president of the Council (©CERN, All rights reserved)

immediate consequences for the price of land. "We didn't even tell the Council at first," said Herwig, "because we were worried that information would leak out."

Having identified where the ring would be placed, the next challenge was surveying, bearing in mind that the geometry of the accelerator could not be adapted to the terrain. It had to be a precise geometrical figure comprising eight circular arcs joined by eight straight sections. The margin for error around the entire 27-km ring was just a few centimetres. Added to that, the tunnel was to be situated as close as possible to the surface, which rises from Lake Geneva towards the Jura mountains. It was therefore designed with a slope of 1.4% to keep it to a depth of 150 metres at most. "Probably for the first time in Europe, a satellite system, NAVSTAR, was used for geodesy," explained Herwig. "It helped us to establish a network of survey points, which was checked continuously for its stability—the survey points were

Fig. 7.10 Herwig welcomes His Holiness the Dalai Lama to CERN in 1983, and receives the traditional gift of a scarf, or khata, symbolising respect (©CERN, All rights reserved)

stable to 2 mm over several years." Another new device, known as a terrameter, which had been developed in the USA for earthquake research, was also deployed in mapping out the future LEP tunnel. It measured distances to the incredible precision of 1 mm over a kilometre by using two laser beams of different wavelengths to eliminate errors due to fluctuations in atmospheric temperature and pressure.

Once LEP had been mapped out on the surface, the reference points had to be transferred underground. "A simple plumb line would not do," recalled Herwig, "we had to take the curvature of the Earth into account since the verticals at opposite sides of the ring were not parallel, but rather converging towards the centre of the Earth." So precise was the model that even distortions to terrestrial gravity caused by the large mass of the Jura were accounted for. Once transferred underground, the tunnelling machines were guided by laser beams. "As a result, the real axis of the tunnel never deviated by more than eight centimetres from the plan," explained

Herwig. "For a 27-km sloping tunnel with just eight access points 3.3 km apart, this is an extraordinary achievement."

With the survey complete, tunnelling could begin. The strategy was to use two tunnel boring machines starting at the lowest point on the ring next to Geneva Airport and progressing uphill in opposite directions around the ring. This would provide safety in case water was encountered, as it would run away from the workface. The more difficult terrain under the Jura had to be excavated by the more laborious process of pilot borings at the workface to investigate what lay beyond, followed by blasting with explosives.

Work began with the access shafts, some of which presented challenges of their own because the molasse is covered by a layer of unstable moraine, which was in places so deep that the ground had to be frozen to $-22\ °C$ before excavation could begin. In the end, the two tunnel boring machines became three, working at different points around the ring. The tunnelling in the molasse went smoothly and was complete by January 1987. Under the Jura, however, it was a different story. In September 1986, high-pressure water broke into the tunnel about 15 m behind the workface, leading to work stopping at that point for eight months. When work resumed, new techniques had been developed to ensure that it would not happen again. "Resin was systematically injected at the front of the tunnel," explained Herwig, "and the tunnel walls were immediately lined by spraying concrete on to them to prevent rockfall and water ingress. The tunnel walls were also strengthened by a metallic

Fig. 7.11 Water ingress under the Jura mountains led to a temporary halt to tunnelling in 1986

vault to resist the water pressure." The delay was partly made up by equipping a short access tunnel leading to the section under the Jura to carry equipment so that tunnelling could continue while the water ingress was being dealt with. Finally, on 8 February 1988, Herwig Schopper and Emilio Picasso found themselves standing deep underground on either side of a thin wall of rock that was all that remained to be excavated. "When the dust from the final blast settled," said Herwig, "we cut a blue ribbon, which was the only obstacle left."

Fig. 7.12 Herwig and Emilio Picasso cut the ribbon at a ceremony to mark the end of the excavation of the LEP tunnel in 1988 (©CERN, All rights reserved)

Fig. 7.13 Herwig Schopper and Abdus Salam present the 1986 Dirac Medal to Yoichiro Nambu at the International Centre for Theoretical Physics (ICTP) in Trieste on 23 July 1987 (© ICTP Photo Archives/Ludovico Scrobogna)

Building the Machine

It's tempting to think of LEP as a rather conventional machine, a scaled-up DORIS or PETRA, but the reality is very different. LEP posed many significant technical challenges, and generated a lot of innovation in accelerator technology, from concrete magnets to superconducting RF and vacuum technologies.

Perhaps the most surprising innovation came in the form of LEP's 3392 concrete bending magnets. Traditional electromagnets consist of iron cores with coils of wire to produce a magnetic field. The sheer size of LEP made this approach financially out of the question, and a more economical solution was needed. The magnetic field required for LEP phase 1 was a modest 0.135 T, and even for LEP II, only double that would be required. That meant that a correspondingly modest amount of iron would be needed, but a very rigid geometrical shape was necessary. The LEP engineers came up with a design that consisted of 1.5-mm-thick low-carbon steel laminations separated by 4 mm and encased in concrete. Rather than using coils, aluminium conductor bars were threaded through the C-shaped concrete yokes to produce the magnetic field. Nothing like this had ever been done before. LEP's concrete magnets served the machine well throughout its operational lifetime, for an investment of about half what conventional electromagnets would have cost.

LEP's vacuum system was another first, allowing a pressure of under a million millionth of an atmosphere, which in turn allowed beams to circulate for several hours before the machine needed to be refilled. The secret was a technique called getter pumping, which was deployed at LEP in the form of a thin ribbon of zirconium–aluminium alloy running around the ring in a channel inside the machine's vacuum chamber adjacent to the pipe in which the beams circulated. This alloy worked like fly paper for molecules, so when all the air had been pumped out using conventional vacuum pumps, the remaining air molecules would stick to the ribbon. Periodically heating the ribbon to 400 °C would free the trapped molecules, reconditioning the getter for future use.

Another of the most challenging components of LEP was the RF systems that accelerated the beams. "In the first phase of LEP, copper cavities had to be used since no other technology was available," said Herwig. But LEP's copper cavities were anything but conventional. RF systems are big consumers of energy. "The electricity cost for the accelerating system turned out to be the dominant part of the operating cost of LEP," Herwig continued. But LEP's design offered a solution to minimise the cost. The machine had been designed to operate with four bunches of particles circulating in each direction, so the cavities would be idle for much of the time and a technique was developed to reduce their energy consumption when they were not actually accelerating beams. "An accelerating cavity consists of several cells in which electric fields oscillate at radio frequencies in such a way that the field points in the right direction to accelerate the particles at the moment a bunch arrives at the cell," explained Herwig. "The fields are produced by electric currents flowing in the walls of the cavities, and even for copper, these currents produce considerable thermal losses." LEP's engineers designed a system in which the field would be transferred to the most energy-efficient structure available, a sphere, when it was not accelerating beams. This led to the striking shape of LEP's copper cavities—a cylinder with a sphere on top, and it allowed the electricity bill for the RF system to be reduced by 40%. LEP's phase 1 RF system consisted of some 128 cavities covering a total distance of 300 metres and installed in two of the tunnel's straight sections.

All in all, the 60,000 tonne LEP accelerator consisted of thousands of components, all of which had to be lowered into the tunnel via a few narrow shafts and delicately manoeuvred into place around the ring. An overhead monorail was installed in the tunnel to transport people and equipment, and on 4 June 1987, the first magnet was ceremonially installed by French Prime Minister, Jacques Chirac, and Swiss President, Pierre Aubert. "A crash programme of installation got underway in September," said Herwig. "As soon as part of the tunnel had been vacated by the civil engineering crews, the installation of components started." Just under two years later, the giant machine was ready to be brought to life.

The Decisions Were Made at the Bar

As LEP had been approved with just a token budget for the particle detectors that would record the machine's electron–positron collisions, a new way had to be found to build these devices. The two big experiments at the SPS collider, UA1 and UA2, set the example to a certain extent, but the LEP detectors would take the new model further still.

The first step to forming the large collaborations that would run the LEP experiments took place at an International Conference on Experimentation at LEP held in Uppsala, Sweden, in June 1980. Jointly organised by the European Committee for Future Accelerators and Uppsala University, this initial conference was followed by a series of ECFA workshops and culminated 12 months later in the Swiss alps, at Villars-sur-Ollon, in a Club Med resort hotel. "The *gentils organisateurs* seemed disappointed that all we wanted to do was sit at the bar and talk physics instead of taking part in all the activities they were putting on," recalled Herwig. "One of my Danish colleagues, Hans Bøggild, summed up the atmosphere in verse: *There once was a place called Villars, A palace with more than one star, They talked about LEP, The future of HEP, But decisions were made at the bar.*"

The decisions were momentous ones, laying down the procedure for forming collaborations and approving proposals for detectors. "My aspiration was to enable as many scientists as possible to do research at LEP," recalled Herwig. "I said that the detectors would be considered as facilities able to perform many experiments and put forward the analogy that the large collaborations were more like a flotilla of small ships rather than a single big ship with one captain. Over time, these small units would combine to carry out common tasks but go their own way to pursue their particular research interests. Some might even break off to join a different flotilla. By and large, this is what happened."

At Villars, it was decided that four of the potential eight interaction regions would be used for detectors, and that letters of intent should be submitted to a LEP Experiments Committee, which would be established in 1982. Proposals would be evaluated on their scientific merit, technical and financial feasibility, and the strength of the collaborations submitting them. Approvals were expected to come in 1983, leaving four years for construction. Crucially, groups from non-member states of CERN would be welcome to take part.

The size of the collaborations that came together following the Villars meeting was unprecedented, with the proposals listing dozens of institutes and hundreds of individuals. They were far bigger than the UA1 and UA2 collaborations. "I also raised a problem of some sociological importance at Villars," recalled Herwig. "The large number of names that would be appearing on the papers would make it hard to appreciate any individual's contribution, so I suggested that common papers about the design and construction should include all names, but specific analyses only those of the people doing the analysis, or alternatively, those names should come first instead of just listing people alphabetically. I completely failed to gain acceptance of this proposal, and the issue is still with us today."

Fig. 7.14 Herwig adopts a suitably solemn expression during a visit to CERN by His Holiness, Pope John Paul II in 1982 (©CERN, All rights reserved)

"In March 1982, the LEP Experiments Committee, very competently chaired by Günther Wolf, began its work evaluating the six proposals that had been submitted," said Herwig. "They were named ALEPH, DELPHI, ELECTRA, L3, LOGIC and OPAL." For the next four months the committee examined every aspect of these proposals, and in July, put forward ALEPH, DELPHI, L3 and OPAL for conditional approval. "ELECTRA and OPAL were very similar, as were DELPHI and LOGIC," explained Herwig, "and we had to ensure diversity among the experiments approved. I was against a brutal rejection of the refused experiments, preferring to give their proponents a fair chance to join the approved ones. In an earlier open meeting of the committee, I'd warned them that the final judgement would not be based only on a strict evaluation of the proposals, but also on factors like degree of technical risk and geographical diversity among member and non-member states. Someone remarked that the committee looks at the science while the Director-General takes care of the politics, and that's pretty much the way it was."

Herwig also had another reason for approving the four experiments, but he kept it to himself at the time. "It was the first time that such complicated and expensive collaborations were tried without a legally responsible officer," he explained. "Two

Fig. 7.15 The support tube at the heart of the L3 experiment was among the more imposing components of CERN's flagship project for the 1990s ()

of the collaborations were headed by strong individuals, while the other two were organised in a more democratic way. I was not sure which model would work best, so I wanted two of each."

A decade later, the procedure for international collaboration that was developed from Uppsala to Villars went on to become the model for the experiments at the LHC, but with one important difference. "To my surprise both models of collaboration at LEP worked very well," explained Herwig, "But it is interesting to note and, worthy of a sociological study, that when the LHC experiments were established, the democratic model was preferred."

The scientific rationale for the approval was that two detectors, ALEPH and OPAL, were general-purpose devices, with OPAL employing tried and tested technology while ALEPH put forward new techniques. DELPHI and L3 were more specialised, each with a different focus. Geographically, they brought together over 100 institutions and some 1000 researchers, mostly from the CERN member states, but also from Canada, China, East Germany, Hungary, Israel, India, Japan, Poland, the USA and the USSR. The committee asked for some changes to be made, and after these were done, conditional approval was given on 18 November 1982.

Before LEP, CERN experiments had all been named for their location and number, hence underground area (UA) 1 and 2, or north area (NA) 1 and so on. Sometimes, the collaborating institutes would create acronyms from their names, hence CERN, Dortmund, Heidelberg, Saclay became CDHS, but when Warsaw joined the collaboration, the name didn't change. "When I was at DESY, I liked the tradition of using acronyms for machines and detectors, so I tried to encourage that at LEP, with mixed success," he recalled. ALEPH, contracted from Apparatus for LEp Physics; DELPHI, the DEtector with Lepton Photon and Hadron Identification; and the Omni-Purpose-Apparatus at LEP, OPAL, all went along with Herwig's request, but the collaboration behind the third proposal to be submitted remained simply the L3 collaboration. It was led by Nobel laureate, Sam Ting. "When I asked Samuel Ting to propose an attractive name, he suggested SAM," said Herwig. "I told him it was not very elegant to use his first name for the detector, but he replied that it had nothing to do with that, SAM stood for "Schopper Approves Me." A few weeks later, he came back with the name Magellan, but the collaboration voted that down on the basis that Magellan, when discovering the Philippines, had been shipwrecked and slain by the natives, which was hardly a good omen for an international collaboration. L3 remained L3."

Slowly but surely, components began to arrive at CERN and the four detectors started to take shape in their caverns. "It always seemed to me like a little miracle that all the parts of these very sophisticated detectors arrived from all corners of the world, fitted together and worked," said Herwig. "It was a huge challenge to get them built in such a short time. To a large extent it was thanks to Horst Wenninger, who I had appointed as technical coordinator for the experiments, that the thousands of different parts fabricated in many laboratories fitted together with extremely high precision."

By the time the detectors were complete, at least one experiment had already proved to be a success: CERN had successfully changed its business model, with the laboratory itself covering just 15% of the total cost of the detectors. Forty-seven percent came from member state institutes, and the remaining 38% from non-member state institutes. The demographic of CERN had also changed, with increasing numbers of non-member state physicists involved. In 1989, the year LEP started up, the number of American physicists at CERN exceeded the number of CERN member state physicists doing their research in the US for the first time.

An Eye on the Future: Superconducting Cavities and Magnets

By the summer of 1989, LEP was ready to start, and four large collaborations were hungry for data. For much of the decade, CERN had put most of its eggs in the LEP phase 1 basket, but not all. Herwig Schopper also had an eye on the future. "Superconducting cavity development in Europe began in my group at Karlsruhe in the 1960s," said Herwig. "I invited Herbert Lengeler from CERN to come to Karlsruhe and lead the superconducting cavity group, which he did for about two years, and when he went back to CERN, he set up a superconducting RF group there." The group was not able to advance the technology sufficiently for LEP I, but design choices were made to ensure that the superconducting cavities for LEP II would be compatible with the infrastructure of LEP I, so they could simply be exchanged. By the time LEP I had done its job in the mid-1990s, Herwig's mandate as CERN Director-General was long over, but the superconducting RF work his teams had begun in Karlsruhe in the 1960s had paid off: superconducting cavities replaced the copper ones as LEP moved into phase 2.

CERN's longer-term future was always in Herwig's mind. In 1984, a workshop organised by CERN and ECFA in Lausanne and at CERN, examined the possibility of installing a hadron collider in the LEP tunnel. This was not surprising: the decision to increase the LEP circumference to 27-km was mainly taken with a view to installing a proton machine to follow LEP. At the time, the USA was discussing a huge machine called the Superconducting Super Collider (SSC) in order not to lose the lead in particle physics to LEP. The SSC was planned to reach energies much higher than could be achieved by a hadron collider in the LEP tunnel, but that didn't mean that a hadron collider at CERN could not be competitive with the SSC. "In a hadron collider, it's the collisions between quarks that provide the interesting physics," explained Herwig, "and since the energy is shared between the quarks, the higher the intensity of the beams, the greater the number of collisions between high-energy quarks." By designing CERN's hadron collider to have much more intense beams, it could be far more competitive with the SSC than the difference in beam energy might suggest. "It was at Lausanne that these discussions started to happen," said Herwig, "and even though the SSC was never built, it shaped the development of the Large Hadron Collider into the powerful research tool that it is today." Prototyping for the extremely powerful magnets that such a machine would require also got underway on Herwig's watch.

CERN Under the Microscope: The Kendrew and Abragam Committees

Another feature of Herwig Schopper's mandate as CERN Director-General was the intense scrutiny that science budgets in general, and CERN's in particular, were under through the 1980s. Once the CERN Council had made it clear that the laboratory would have to operate under a constant budget, one member state, the UK, went a step further, setting up a committee with Herwig's old friend John Kendrew in the chair. Its remit was to review UK particle physics, and in particular the UK's financial contribution to CERN. One year later, the committee, consisting largely of scientists from fields other than particle physics, delivered its report. While praising the scientific excellence of CERN, it recommended that the UK should remain a CERN member state until LEP was complete, but then withdraw if the UK's subscription could not be reduced by 25%. "It took me some time to understand what was behind this recommendation," explained Herwig. "Unlike many CERN member states, the UK did not have a dedicated budget line for CERN: the UK contribution was channelled through the Science and Engineering Research Council (SERC) instead, so CERN was in direct competition with other research budgets. To exacerbate the situation, the CERN budget had to cover all CERN expenses, not just research, so it took up a large fraction of SERC's overall spending. If the CERN contribution increased because of the Sterling to Swiss Franc exchange rate, the SERC budget suffered. CERN's annual budget is about the same as that of a large university, but SERC did not have to pay all the costs of universities, so this arrangement seemed unfair to me."

It took a direct intervention from the UK prime minister to resolve the situation. "When Margaret Thatcher visited CERN, I had explained these difficulties to her," Herwig recounted. "At the end of her visit, she told journalists that she was convinced that the UK's contribution to CERN was well spent." In 1987, the UK formally decided not to give notice of withdrawal from CERN, but the intense scrutiny was not over. By this time, another committee had been established, this time by the CERN Council itself, but at the instigation of the UK delegation. It was chaired by the renowned French physicist Anatole Abragam and was established in February 1986. Its mandate was to review the cost-effectiveness of CERN's use of human and material resources, along with employment conditions at the laboratory.

The Abragam committee consisted mostly of industrialists and people with experience in modern management methods. One of them was Miguel Boyer, president of *Banco Exterior de España* and a former finance minister. When the committee came together for the first time, Abragam distributed the tasks among the committee members, and CERN's finances went to Boyer. "As a former minister, he was accustomed to a certain lifestyle," recalled Herwig. "So he checked into top hotels and turned up for meetings in a chauffeur-driven limousine. Since CERN had to pay the expenses of the committee, I told Abragam this was not acceptable, and he made it clear that I'd have to talk to Boyer if I wanted things to change. It turned out to be easier than I expected, Boyer understood completely and spent his time in Geneva more modestly after that. A bigger problem was Carlo de Benedetti."

De Benedetti was the CEO of Olivetti. "In the very first meeting," recalled Herwig, "he asked me about the turnover of staff at CERN, which was about 5–6% per year. He took this as meaning that CERN was a sclerotic organisation and told us that Olivetti had a turnover of about 30% per year. This turned out to be due to an early retirement programme, and the committee took great interest in that. They ended up recommending a similar programme to CERN, but unlike Olivetti, which just pushed people into Italy's state pension fund, such a programme would create a financial burden for CERN's pension fund."

As well as recommending an early retirement scheme, the Abragam committee suggested that international status not be granted to non-professional staff, so they would fall into the social security systems of CERN's two host states, the introduction of performance-related pay and promotion, a reduction in the granting of indefinite contracts, and a range of recommendations for the management of CERN's accounts and pension fund. The committee did not recommend, despite the UK delegations request to do so, a reduction of 25% in CERN's overall budget.

"In some shape or form, most of these recommendations were implemented," said Herwig, "leading again to calls for my resignation. When the Council agreed to the early retirement scheme, they did not allocate resources to cover the cost to the pension fund, so that had to be compensated from the already depleted CERN budget. I had no choice."

And the Answer is Three: The First Results from LEP

By July 1988, the first octant of LEP was ready and positrons were injected into the machine for the first time. It was a moment of triumph for CERN, demonstrating that the injector chain, consisting of a new facility to produce and store electrons and positrons, along with the PS and SPS, was working to plan, and that there were no obvious show stoppers with the LEP machine itself. One year later, LEP was complete, and CERN chose 14 July, the bicentenary of the storming of the Bastille, to test the whole machine for the first time. "The control room was packed as the engineers and technicians anxiously watched the first attempt to coax a beam around the ring," recalled Herwig. They needn't have worried—the beam sailed round with barely a hitch, and LEP commissioning was underway. On 25 July, electrons were successfully injected for the first time, and by 12 August, beams of electrons and positrons were circulating in opposite directions around the ring at the injection energy of 20 GeV. The RF accelerating system was turned on to boost the beams up to around 46 GeV. Everything was ready for LEP's first collisions, and all eyes turned to the control rooms of the four big experiments. Early in the evening of 13 August, things were not looking too promising as the beams were lost, but the decision was taken to try again. At 21:43, LEP's operators once again started to accumulate beams. Seen from the experiments' control rooms, progress seemed painstakingly slow, but at 22:52, the monitor screen, known as LEP page 1, displayed the word 'ramp', indicating that the beams were being accelerated. At 23:00, the electrostatic separator plates keeping the beams apart at the collision points were switched off, and 'ramp' was replaced by 'collide'. For 16 agonising minutes, nothing happened, but then on the event display monitor in the OPAL control room, the unmistakable image of a Z-particle decay appeared. "By the time a printed copy of the event had been brought from OPAL to the LEP control room, a further five events had been reported," recalled Herwig. "LEP was underway and a new era in experimental physics had begun."

On that first night, ALEPH and L3 also reported their first collisions, with DELPHI joining the party the following day. Beams had not quite been lined up in DELPHI, but as soon as that was fixed, data started to flow in LEP's most speculative detector

Fig. 7.16 Carlo Rubbia and Herwig Schopper in a crowded LEP control room on 14 July 1989 awaiting the new accelerator's first beam (©CERN, All rights reserved)

Fig. 7.17 Playing the piano at his only public performance—a ceremony to mark 25 years of service for CERN staff in 1988. Helga Schmal, who served as assistant to Directors-General from Willibald Jentschke to Luciano Maiani, turns the pages (©CERN, All rights reserved)

as well. After more fine tuning of the machine, a pilot run began on 20 September and continued to the end of the year.

The first big question that the LEP experiments had to address was the number of generations of fundamental particles that exist. The kind of matter we consider ordinary, which makes up everything visible in the universe, consists of up quarks, down quarks and electrons, along with the neutrino predicted by Wolfgang Pauli in 1930. Cosmic rays had revealed a second generation consisting of strange quarks, charm quarks, muons and muon-neutrinos, and members of a third family—bottom quarks and tau particles—had also been discovered by the time LEP started up. Studying the decay of Z particles offered a way to find out if there were more generations to be discovered, assuming that any as-yet undiscovered neutrino was light enough to be produced by a Z particle decay.

It would be a quick measurement to make, but the LEP experiments had a race on their hands. The Stanford Linear Accelerator Center (SLAC) in California had added bending arcs to the end of its two mile long linac, allowing it to accelerate electrons and positrons and bring them into collision with enough energy to produce

Fig. 7.18 Herwig shows the CERN *Livre d'Or* to husband-and-wife visitors, the author Friedrich Dürrenmat and filmmaker Charlotte Kerr (©CERN, All rights reserved)

Z particles. The Stanford Linear Collider (SLC) had been operating since April, and by the time LEP started up, Stanford physicists were showing results based on a sample of 233 Z particle decays. LEP's collision rate was much higher than that of the SLC, so it was only a matter of time before the LEP experiments would overtake the American collider, but the race to publish the so-called Z line shape was real.

When the collision energy of electrons and positrons reaches the energy of the Z particle, a peak appears in the distribution of events containing particles the Z can decay into. This is called the Z line shape, and it is sensitive to the number of generations of fundamental particles. The detectors at LEP and the SLC can detect all the types of particles that Zs can decay into, except for neutrinos, which escape the detector unseen. That allows physicists to model the line shape for any number of particle generations, assuming that the neutrinos are light enough in every generation to be produced in Z particle decays. By comparing the measured line shape to predictions for two, three or four generations, LEP and SLC physicists could determine how many generations there are. In August 1989, SLAC's sample of 233 events allowed them to say that the upper limit on the number of particle generations was 4.4.

When the LEP pilot run got underway, the LEP experiments were recording thousands of Z particles per day, and they were preparing to present their first results in October. Meanwhile SLAC was homing in on a definitive result, presenting updated results at the European Physical Society's September meeting in Madrid. It seemed almost inevitable that the data would show that the three known generations were all there are, and confirmation was soon delivered. On 13 October 1989, CERN issued a press release with the title "First Physics Results from LEP." It included the news that "there are no other neutrino types in nature beyond the three associated with the electron, muon and tau particles." The final result from LEP, published at the end of LEP I running and combining the results of all four experiments, was that the number of light neutrinos is 2.9840 ± 0.0082. Three, in other words. Why only three kinds of neutrinos exist remains an open question to this day.

Herwig Schopper's mandate as Director-General of CERN was extended beyond the initial five-year term, but had nevertheless ended on 31 December 1988 when he passed the baton to Carlo Rubbia. When the machine was formally inaugurated on 13 November the following year in the presence of heads of state and government from CERN's member states, it was therefore Emilio Picasso who had the job of handing over the symbolic keys to the machine to the new Director-General. Although Herwig had no part to play in the ceremony, it was very much a celebration of his mandate. Not only had CERN succeeded in constructing the world's largest scientific instrument under particularly difficult financial conditions, but there were also 14 countries represented on the podium with Spain and Portugal having become member states during Herwig's mandate.

Fig. 7.19 The LEP experiments' measurement of the hadron production cross-section around the Z particle resonance. The curves show the predicted cross-section for two, three and four neutrino species with Standard Model couplings and negligible mass. This measurement clearly shows that there are three, and only three, types of neutrino (©CERN, All rights reserved)

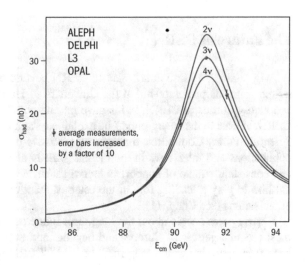

Fig. 7.20 In 1996, L3's spokesperson, Sam Ting (left), shows a delegation from the US Congress around the experiment. Ting's colleague, Jim Allaby, is to the right (©CERN, All rights reserved)

In February 1989, Herwig Schopper turned 65. After an illustrious career he had earned his retirement, but Herwig had other ideas. He stayed on at CERN to re-join the ranks of particle physicists, becoming a member of Sam Ting's L3 experiment and contributing to the publication of several papers. Then it was off to UNESCO and other tasks.

The Future of CERN

Herwig retained a close attachment to CERN beyond his retirement. Having also retired from his professorship at the University of Hamburg, he stayed in Geneva and kept an office at CERN. "This gave me the possibility to continue to support CERN, at least in the background," he explained, "as a former member of the CERN Scientific Policy Committee, through my extensive personal contacts, and through publications and interviews in the media." CERN also proved to be a very strong and consistent source of support to Herwig later, when his post-retirement career led him to play a leading role in the establishment of new scientific international organisations (see Chap. 11).

Herwig cares deeply about the long-term future of CERN. "CERN is one of the most precious jewels of Europe, and beyond, and every effort should be made to keep it in operation in the long run," he said. "CERN is a symbol with respect to carrying out projects on time and within budgets. It has a very diverse programme, but always with big flagship projects that are outstanding on a global level. The present Director-General, Fabiola Gianotti, is following this tradition as an advocate for a large tunnel with a circumference of nearly 100 km, which, like the current tunnel, would host first an electron machine and then a proton one."

Whether this or another project is ultimately pursued, Herwig believes that Europe should not abandon the model for international research that CERN has honed over the decades. "Such a project should not be built for national or regional prestige," he said, "but rather in a spirit of world-wide cooperation."

With his long career in research, stretching back to a time when Europe was emerging from a brutal period of conflict, Herwig is always mindful that CERN has a second strand to its mission. "In the present political situation," he explained, "one of the original ideas of CERN could become important again: science for peace. CERN was founded with the idea of helping to bring European countries together in peaceful cooperation, and it has contributed to this aim considerably." CERN's neutrality even played a role during the disarmament talks in Geneva in 1985. "In the wake of the Reagan–Gorbachev summit of November 1985, disarmament negotiations seemed to be getting nowhere," recalled Herwig. "One day the head of the US delegation, Alvin Trivelpiece, a physicist that I knew, called me to ask if I would invite the heads of the two delegations to dinner at CERN, where the neutral relaxed atmosphere might lead to a breakthrough. So I did, and that seemed to get things moving in the right direction."

Herwig's vision is that the CERN model could be applied in other fields, uniting humanity around common goals. "CERN's main product is the increase of fundamental knowledge concerning the laws of nature," said Herwig. "Fundamental research is independent of any economic, political or military interests, and this may be true for other fields as well. As a result of the CERN model, no other field is better integrated than particle physics into the cultural environment of so many countries around the world. I can think of many areas of research that would benefit from such an approach. I hope that for the future of CERN, and also research more generally, we can rekindle the spark exemplified by Denis de Rougemont and his 1948 European Cultural Conference. This would be my dream! CERN and science certainly cannot help to solve political crises. These are usually resolved by peace treaties or at least armistices. But such agreements are not worth the paper they are written on unless they are based on a minimum of mutual trust. The past has shown that science is a wonderful tool that can help create such confidence."

Fig. 7.21 With Queen Beatrix of the Netherlands and Prince Klaus at CERN in 1985 (©CERN, All rights reserved)

In His Own Words: Encounters with Remarkable People

A Prime Ministerial Visit

"In August 1982, we had a visit from Margaret Thatcher, she came to CERN because she was very interested in science. She introduced herself by stating that she did not want to be dealt with as a Prime Minister, but as a fellow scientist. She had a degree

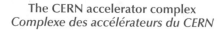

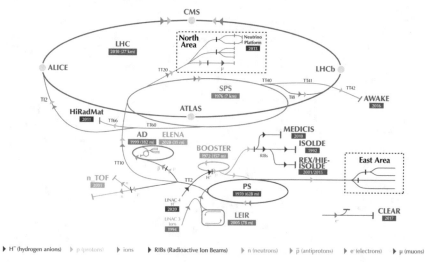

Fig. 7.22 The CERN accelerator complex in 2022 showing the diversity and complexity of the Laboratory's unique collection of research facilities. CERN's current flagship, the LHC, occupies the tunnel constructed for LEP during Herwig's mandate (©CERN, All rights reserved)

in chemistry. At lunch, she asked me the question, 'What will be the size of the next project after LEP at CERN?' I said, 'There will be no bigger ring at CERN because there is no more space. LEP is the last of the big rings at CERN.' Then she answered, 'I asked the same question of John Adams when he built the SPS and he gave me the same answer, why should I believe you any more than him?' Now many years later CERN is discussing a new tunnel with a circumference of about 100 km. One should never try to predict the future, because one never knows what might happen!

During her visit I introduced Mrs Thatcher to Carlo Rubbia. In August both experiments at the SPS collider, UA1 and UA2, were starting to see indications compatible with a W particle, but the uncertainty was still very large. After talking to Carlo, she was fascinated by the possible discovery, and she said to me, 'Well, Mr Director-General, I do not want to learn about the discovery from the press. I want to be informed before you go to the press.' I promised her that I would do this.

Then, when Carlo Rubbia and Pierre Darriulat showed me their data, I was afraid that rumours of this discovery would leak out to the press and Mrs Thatcher would be very angry for not having been informed before. That was just before Christmas 1982, and in January '83 there was a big international conference in Japan that I had to attend, so before I left CERN I called Rubbia and Darriulat and told them the story

Fig. 7.23 With British Prime Minister Margaret Thatcher and husband Denis to her left at CERN in 1982 (©CERN, All rights reserved)

of Mrs Thatcher, that she wanted to be informed early, and I said to them, 'if you want to go public in January while I am in Japan, fine, you can go ahead, but before you do so, you have to send me a fax to Tokyo so that I can inform Mrs Thatcher that the announcement is imminent.' Just to be sure that I would keep my promise, I wrote her a letter on 20 December saying that the discovery appeared to be imminent.

Soon after I got to Japan, I received a fax from Rubbia and Darriulat, email was not much used at the time, and I sent a fax to Mrs Thatcher, asking her to keep the news confidential until it was officially announced. She replied immediately, saying, 'You can be sure I'll keep it in strict confidence before it's published.' And a few days later there was a colloquium at CERN where the discovery was announced.

During her visit to CERN, Mrs Thatcher asked many pertinent questions demonstrating a real interest in physics. She was accompanied by her husband, and throughout the visit I was impressed at the tenderness between the two of them.

She may have been a very busy Prime Minister, but she clearly did not overlook the importance of her personal life."

An Interesting Collaboration

"Among the many remarkable people I had the opportunity to know over my career is Antonino Zichichi. Nino to his friends. Nino is an excellent physicist and was employed by CERN soon after its foundation. It's there that he carried out most of his research. I remember him having boundless energy and drive, which allowed him to achieve much beyond his CERN activities.

Antonino Zichichi was born in Trapani and educated in Palermo. He was always deeply rooted in Sicily. He once told me that the Sicilian mafia originated as a political organisation to liberate Sicily from foreign overlords. He also had some family connections to the catholic church in Sicily, and even to the Vatican. With this background Nino established a scientific teaching centre in the 1960s in Erice, a small, isolated town on a steep conical hill above Trapani. He called it the Ettore Majorana Foundation and Centre for Scientific Culture [1]. Erice is a beautiful place with a long history. There are stone plates there with Phoenician inscriptions, and during the Middle Ages it became the seat of several small monasteries. By the middle of the twentieth century, these were mostly abandoned and decaying. Nino somehow managed to get hold of a few buildings and he converted some of the rooms into lecture theatres where small workshops and courses could be organised. With CERN's support, the first ones dealt with nuclear and particle physics. Some of the monastic cells were converted into simple guest rooms for the participants. At one point when I was still at Karlsruhe, Nino asked me to help to stabilise the enterprise. I agreed to take the responsibility for organising courses for medium-energy nuclear physics. Although the accommodations were quite rudimentary back then, the stay at Erice was quite charming. There was a beautiful view over the Mediterranean, and we could eat in one of the town's many restaurants. The whole setting is quiet and peaceful, and very conducive to learning. Originally the courses were organised during the academic vacations in summer, but as the centre's success grew, courses were extended to other fields of physics and beyond. Today, there are schools in Erice all year round. Most of the old monasteries are now integrated into the centre, which has become an important economic element of the town. The lecture halls are now perfect, the guest rooms quite comfortable and Nino has a grandiose office in one of the old towers dominating the whole scene. Nino, who has a good sense for publicity, named the centre after Ettore Majorana, who was a young Sicilian theoretical physicist who mysteriously disappeared in 1938. Over the years, Nino invited me to attend all kinds of interesting conferences, often involving Nobel Prize winners, and I had the opportunity to explore the beautiful island of Sicily.

Immediately after my retirement as Director-General of CERN, Nino asked me to help him with the 'World Laboratory,' which he'd set up in Lausanne in the 1980s. Although the title seemed to me somewhat pretentious, I agreed to become the chair of its management committee. Nino had obtained a budget of several million Swiss Francs from the Italian government to finance various projects in physics or other scientific domains in developed and underdeveloped countries. For a year I went almost daily to Lausanne. The administration was kept to a minimum and consisted of a professional manager, clerical staff and me with Nino as the President, of course. We managed to get things off the ground but after a bit more than a year the load became too heavy, and I stopped. I don't know what became of the World Laboratory after that."

Reference

1.   https://ettoremajoranafoundation.it/the-history/

Chapter 8
From Science to Science Diplomacy

After retiring from his career in science, Herwig embarked on a new one as a diplomat. "Looking back at my professional life," he explained, "it can be divided into three parts: first I was a physicist, then a manager of laboratories, and the third part is as a kind of diplomat." It could be argued that Herwig's entire career had been leading towards science diplomacy, and in retirement, that is indeed the direction he took.

When Herwig reached his 65th birthday on 28 February 1989, his professorship at the University of Hamburg became, following the time-honoured tradition of academia the world over, and in accordance with German law, an emeritus position. On the same day, his position at CERN came to an end, as did the leave of absence from Hamburg that had allowed him to take the role. "On the one hand, I enjoyed a smooth transition from active employment to retirement, and I could easily have returned to Hamburg," he recalled, "but on the other hand, by this time, my home had become Geneva. I had no close relatives any more apart from my immediate family, and my children had made a home in Switzerland. Ingeborg and I decided to stay."

It was not long until other organisations began to seek Herwig out. First to call was the DPG, the German Physical Society, or *Deutsche Physikalische Gesellschaft*, for which he served as president from 1992 to 1994. The DPG is the largest physical society in the world with about 50,000 members, many of them young physicists: something the society prides itself on. Its seat is at Bad Honnef near Bonn. As with any learned society, the DPG organises conferences, provides support to its community, and nurtures links with schools and industry. The usual role of the president is to chair the governing board and to oil the wheels of the society's activities, assisted by a permanent staff. Herwig, however, had a bigger task to accomplish. "The DPG always gets the most it can out of its presidents," he explained. "Each one is active for many years, because as well as being president, they also have duties as president-elect, and as immediate past president. When I was elected, a historical event was underway—German reunification—and that also applied to the two German physical societies. I oversaw the conversion of the former headquarters of the East German

© The Author(s) 2024
H. Schopper and J. Gillies, *Herwig Schopper*, Springer Biographies,
https://doi.org/10.1007/978-3-031-51042-7_8

society into a new representation of the DPG in Berlin. It was in the Magnus-Haus, one of the oldest buildings in Berlin, and this was a particularly interesting task."

This presidency was followed by the presidency of the European Physical Society (EPS), which Herwig held from 1995 to 1997, again with responsibilities as president-elect and immediate past president. "This was also quite challenging," he recalled. "The society had been in Geneva since it was founded but it was facing severe financial difficulties, so I moved the headquarters to Mulhouse, which was more affordable for the society's limited resources." Herwig's tenure also saw an increased opening to Eastern Europe as it emerged from behind the Iron Curtain, and in particular to Hungary. "I strengthened the EPS office in Budapest," he said, "this was one of the very rare links between East and West at the time. The Hungarian physicist Norbert Kroó, who had preceded me as EPS president, provided much support. My task was very much supported by Márta Neményi, a scientific secretary at the Hungarian Academy of Sciences. I spent a lot of time there, which allowed me to discover the beautiful country of Hungary."

There is one more activity that, unknown to Herwig at the time, allowed him to develop the skills he'd need for the third phase of his career. For several decades, he had been an editor for Springer's venerable series of data collections, known simply by the name of its founders: Landolt Börnstein. "Looking back," he recalled, "I do not understand how I could do this on top of everything else, with only 24 hours a day available."

Nevertheless, Herwig had been doing his part to ensure that the famous articles and data tables remained up to date, relevant and useful to scientists for much of his professional life. "As a young scientist, I contributed articles in optics and nuclear physics," he said. "Later I became editor for the nuclear and particle physics section, and finally for the whole of physics." Under Herwig's tenure, over 30 volumes were published with Herwig as editor or co-editor. "The changes in publishing over the years have been remarkable," he said. "At the beginning, manuscripts had to be submitted in multiple typed copies whereas today the whole process is digital. I'm also a bit proud that I introduced some special review volumes for particular topics, including three for particle physics, which have recently been published open access under the CERN umbrella."

Through all these activities, Herwig was honing his diplomatic skills, but it was a call from UNESCO, however, that brought everything together, and opened the way to the third phase of Herwig's career. He was about to become a science diplomat.

ROSTE—Getting to Know Venice

In the late 1980s, UNESCO had set up a body known as the Regional Office for Science and Technology for Europe (ROSTE) in Venice [1]. Herwig's successor as Director-General of CERN, Carlo Rubbia, was a member of the office's first Scientific Council, which was set up in 1989. The date was no accident. By the late 1980s, the post-war order in Europe was crumbling, and the Italian government made an offer to

UNESCO to transfer its existing regional bureau for scientific cooperation in Europe from Paris to Venice [2]. "UNESCO wanted to support countries on both sides of the crumbling Iron Curtain," explained Herwig, "and the Italian government had in mind to do something similar for the region of Friuli Venezia Giulia, and neighbouring countries in South East Europe." ROSTE's Scientific Council advised the office's governing body. "The activities were controlled by a supervisory committee with two delegates from Italy and two from UNESCO," explained Herwig, "they took the final decisions on how to use the money available, but they were advised by the council."

By the time Herwig was approached to join the ROSTE Scientific Council in the late 1990s, the office had developed a broad programme of activity covering basic and applied science, as well as societal issues such as addressing brain-drains. It was also active in the application of science and technology to the conservation of cultural heritage, in which the region is rich. "The office was installed first in a beautiful old palazzo located right on the waterfront very close to the famous wooden bridge leading to the museum of the Accademia," recalled Herwig. "These old palazzi belonged to the rich Venetian families that made their fortunes by trading with the Far East or the eastern Mediterranean. The ground floors, close to the water, were used to store goods for ease of access to transportation. The family lived on the first floor. The main floor, and the upper floors were used for offices and to house servants. Over time, these families became poorer, so they could no-longer afford to maintain these palazzi, so they rented the first floor to organisations and installed themselves in the upper floors which were originally used by the servants. So ROSTE was installed on the first floor of the palazzo with a beautiful view of the lagoon from the balcony. It was a real pleasure."

Beautiful though it was, the living quarters of a wealthy Venetian merchant family were not up to the demands of a modern office, so towards the end of Herwig's time with ROSTE, UNESCO's Venice office moved to the Palazzo Zorzi, which had been refurbished to meet modern office standards. "It was very nice from the technical point of view, but one disadvantage was that it was on the ground floor," recalled Herwig. "When there was a bad winter and Venice was flooded, the so called *acqua alta*, then the ground floor of that building was flooded too. I remember sometimes we got into the meeting rooms only by putting on high boots and walking through water some 10 or 20 cm deep."

Herwig served as a member of the ROSTE Scientific Council from 1998 to 2002, for the last two years as its president. During this time, he was able to develop further contacts that he'd established while at the European Physical Society with Eastern European countries. "In a way, it was a continuation of my previous activities," he said. But there were differences too, since because of the Italian government's support, ROSTE also had a mandate in the Friuli Venezia Giulia region of the country. "Sometimes, completely different tasks were involved," he continued. Part of ROSTE's mandate was to help deal with environmental issues such as the pollution of Venice's waterways. "This was a complicated matter of planning and management," he recalled, "but there were more enjoyable aspects. We got involved in supporting the opera house at Padua, because ROSTE had evolved to support

Fig. 8.1 The magnificent Palazzo Loredan dell'Ambasciatore, so called because it was used by ambassadors from the Austrian Empire, served as a UNESCO office when Herwig began his time at ROSTE. Meetings took place in the room with a balcony overlooking the Grand Canal (©Didier Descouens, Wikimedia Commons, CC BY-SA 4.0)

cultural activities as well as science. This was a very interesting part of my time with ROSTE."

The ROSTE Scientific Council met every two to three months in Venice, giving Herwig the opportunity to gain a deeper insight into this unique city than any tourist every could. "There were outstanding European scientists in the council, including, of course, Italian colleagues, some of them from Venice," he explained. "They knew the town and its history very well, and sometimes in the evening after the meetings, they showed us Venice and explained the history and secrets of the town. So, I got to know Venice very well, and knew all the tricks: for instance, I very much liked to visit the cathedral, which is full of treasures, but there's usually a long queue of tourists and it's difficult to get in." Thanks to his Venetian colleagues, Herwig discovered the secret entrance. "There was a little side door that was always open for those who wanted to pray, and I sometimes used that when I didn't have much time, just to get a glimpse of the treasures inside."

One of ROSTE's meetings fell within Venice's famous carnival, which was an eye opener to Herwig. He was used to the raucous nature of the Mainz carnival, and he discovered that Venice's version was very different. "In Germany, carnival is a very happy and joyous event, but not so in Venice," he began. "In Venice everybody is masked, but it's not just the face, but the whole body that's disguised, so you can't even tell who is a man and who is a woman. Everyone is anonymous: they don't even speak. Carnival in Venice was completely different to what I'd expected, and I also learned from my colleagues that in the old times the carnival played an important social role because for a few weeks all people were equal."

The UNESCO International Basic Sciences Programme

In 2002, Herwig left ROSTE, but his most important work with UNESCO was still to come. "The 'S' of UNESCO stands for science, so UNESCO in a way was responsible for supporting and taking care of science," he explained. "It was very early in the 1990s, I think, that a committee was set up to support the creation and management of large facilities." Forty years earlier, CERN had been the first large research infrastructure to see the light of day under the auspices of UNESCO, and with the success of that endeavour clearly evident by the 1990s, why not try to replicate it in other regions, and in other fields of science?

The year before Herwig left ROSTE, the eminent Polish biochemist, Maciej Nałęcz, was appointed director of UNESCO's division for basic and engineering sciences, and he set about establishing an international basic sciences programme (IBSP) to promote and support fundamental research. "Nałęcz is really a very nice man, fair and impartial, very much dedicated," recalled Herwig. "UNESCO support for science depended on his effort and engagement for many years, and I became very close to him professionally, and also in private life."

By 2003, the IBSP was up and running and Herwig had agreed to be its chair, a position he held until 2009. "IBSP was a very interesting committee because there was strong interest from Russia, notably a former minister, Mr. Fortov, and also important people from Africa," recalled Herwig. "I remember one person from Rwanda, Romain Murenzi, who became a government minister, and later director of the Third World Academy of Sciences in Trieste."

Among the programmes proposed for the IBSP was one that was favoured by Fortov. "The Russians had already rolled out a prototype of a very small and cheap satellite and the idea was that many of them could be used to bring teaching directly into developing-world classrooms, but unfortunately it never took off," explained Herwig. "Another possibility was to teach school teachers." Although the initiative itself did not succeed, it inspired Herwig to pursue the same goal through other channels. "I proposed a collaboration between this IBSP programme and CERN," he explained, "because CERN already had a programme to teach and train

Fig. 8.2 Russian Vice-Premier Vladimir Fortov presents an Order of Friendship to Herwig during a session of the JINR Scientific Council in January 1997 (©CERN, All rights reserved)

schoolteachers. The particularly interesting part of the programme, which covered mainly physics and related fields at CERN, was that the teachers got courses in their own language. CERN could only organise programmes for countries that were members of CERN, so I proposed that this programme should be extended to other countries, in particular to Africa, with the help of UNESCO, and that worked quite well." The programme received funding from the UNESCO IBSP for several years. "Unfortunately, I must say, when the DG of UNESCO changed and Mrs. Bokova became the new DG, the situation became very difficult," said Herwig. "The US decided to leave because UNESCO had agreed to recognise Palestine as a country, whereas in the UN, that was not the case and America strongly supported the Israeli position."

Before pulling out, the US had provided about one-third of the UNESCO budget. "The result was that UNESCO almost completely lost its interest in science and neglected the 'S' in its acronym," Herwig said with regret. "I still had several meetings with Director-General Bokova. I asked for help for SESAME at that time but without result. It seems that recently, UNESCO is changing its policy again. With the new Director-General, maybe UNESCO is getting more interested in science again, but unfortunately the US has not yet come back, and without money not much can be done."

The Nobel Prize-Winner and the President

Herwig's involvement with UNESCO led to some unforeseen consequences, including an encounter between a renowned physicist and a president with his eye on a third term in office. "James Cronin, a very well-known physicist who got the

Nobel Prize, spent a sabbatical year at CERN in '82–'83 and we got to know each other quite well, recalled Herwig. "Several years later he became one of the main promotors of an international project for astrophysics." The idea was to create an observatory to study the highest energy cosmic ray particles striking the earth. Such an observatory needs to cover a very large area since the highest energy cosmic rays are very rare, so to catch enough to do meaningful research, the detector array has to be big. The observatory was eventually built in Argentina, where an ideal location was found in the province of Mendoza. It's called the Pierre Auger Observatory after the French physicist who was also one of the founding fathers of CERN.

The Pierre Auger Observatory was proposed by Cronin along with the British physicist Alan Watson in 1992. By the late 1990s, when Herwig was established at UNESCO, the site had been chosen and a management structure was being set up. Herwig was invited to chair a finance board. He agreed to chair a meeting at UNESCO headquarters in Paris because it appeared to offer a way to help his friend Jim Cronin, although other commitments prevented him from going on to chair the board long-term.

Argentina's president, Carlos Menem, was scheduled to make an official visit to France in October 1998, so on the 13th of that month, a meeting of the Pierre Auger Observatory's finance board was convened at UNESCO. "It's not usual for a head of state to visit an international organisation while on an official visit to a country," said Herwig, "but they could choose for personal reasons to attend a meeting hosted by an international project, so we invited Menem." The president accepted and was invited to speak along with the DG of UNESCO. "I thought it was quite amusing that I had to give the floor to the DG of UNESCO in his own house," said Herwig, "but those are the rules of protocol."

President Menem used the occasion to make the announcement that construction of the Pierre Auger Observatory would begin the following year. Although his ambition to run for a third term came to nought, the Argentinian constitution forebade it - Menem was good to his word. The Pierre Auger Observatory formally came into being in March 1999. Construction began two months later and in 2009, the detector array, covering an area similar to that of Luxembourg, was formally inaugurated [3].

"I get a little bit frustrated sometimes with all the stories about creating new projects and facilities," recalled Herwig. "So much of the story is lost. All you find is what's in the minutes of meetings, but in the minutes only the formal decisions are recorded. The real motivation of people, how things were organised and how the ideas evolved, are not in the official minutes. Today, if you look up the Pierre Auger Observatory, which has become a very successful international project, so much of its history is forgotten. You find the name of Jim Cronin, of course, but there's no real understanding of the amount of work that he put in to getting the project going, and that's a little bit sad."

Excursions into Nuclear Fusion

"Looking back at my life, I don't know how I did all these things after I retired from CERN, because the day was still only so long, and I had still a family and other interests," said Herwig. "I was combining very many activities in parallel, with a lot at the same time."

One of those activities goes back to the early part of Herwig's career, when he was working in Hamburg. "In the 1970s physicists in Europe, or even in the whole world, were still a relatively small family, we were not so many people like today, so we knew each other quite well," he explained. "Nowadays it's difficult to get to know all the colleagues even in your own field." In the 1970s, the disciplines of physics were also much more fluid than they are today. "I must say, I preferred that because I was always interested in physics as a whole." As a consequence, Herwig came to know astrophysicist, Arnulf Schlüter, who moved from theoretical astrophysics in 1965 to become director of the Institute for Plasma Physics (IPP) in Munich. "This institute was founded at the initiative of Heisenberg," said Herwig. "If we look back, it's always the same few people with vision asking questions about natural phenomena, so it was Heisenberg who proposed the institute in Munich." Established as an independent institute in 1960 in Garching, just outside the Bavarian capital, it soon became part of the Max Planck Society and remains so to this day.

Schlüter wanted to understand the fourth state of matter, which makes up the bulk of the universe. "The original idea in Munich was to study plasma physics," explained Herwig. "Plasma is a heated gas, heated to such an extent that the atoms became ionised, which means they are now charged particles, and therefore behave completely differently to a normal gas, because the particles of the gas feel an electromagnetic interaction. The behaviour of plasma clouds in the universe is completely different from other dust or gas clouds, so many people were studying the behaviour of such plasma clouds in the universe."

Stars are made of plasma, and they generate energy through nuclear fusion reactions in the extremely hot, dense plasmas at their cores. By the 1950s, people had started to speculate whether it might be possible to replicate this process on Earth. Eminent physicists such as Andrei Sakharov and Igor Tamm [4] in the Soviet Union proposed a device called a tokamak in 1950, which would confine super-hot plasma inside a magnetic field where the conditions for fusion would be right. One year later, an alternative magnetic confinement device, called a stellarator, was proposed by the American theoretical physicist Lyman Spitzer. Herwig became aware of magnetic confinement fusion energy research at the second Atoms for Peace conference held in Geneva in October 1958, where the Russians revealed their work on tokamaks for the first time to the international community.

"In a way, it's the cleanest way of producing energy because there's no big problem with waste," explained Herwig. "But the question is, how do you get it working? Of course, one could accelerate protons to high energies in an accelerator and shoot them at other protons, but the number of fusions would be much too low. Another possibility is to copy the Sun by heating the plasma to a very high temperature so that the motion of the individual nuclei becomes so fast that they overcome the electrostatic repulsion, and fuse together. The problem then is that when the plasma is very hot, it's very difficult to enclose it because it would melt any material you tried to hold it in. The only way is to enclose the plasma by magnetic fields, but that's easier said than done." Magnetic fields keep the plasma away from the walls, but the plasma is very unstable, breaking up easily and losing energy. "Understanding plasma instabilities, and how to control them soon became one of the main topics of plasma physics, not only from a scientific point of view, but also from a practical point of view to create fusion," said Herwig. "Whenever I visited the IPP, I asked Schlüter, "How fast do you think you'll go?" Every time I went there, I asked that same question, and Schlüter always gave the same answer: "It will take 30 years." So, after going there for 10 years and still getting the same answer, I said: "Look, you always give me the same, answer, aren't you making any progress?" He said, "we are very consistent, we always give the same answers, we don't change our position." That was, of course, a joke, but the trouble is, this problem is still a serious issue in fusion energy research even after about 60 years."

Because of his personal visits to the IPP, friendship with Schlüter, and his experience with large projects, Herwig was later invited to become a member of the institute's board of governors, its *Kuratorium*, on which he served for over 12 years. "The Kuratorium is the highest governing body of the IPP and is chaired by the president of the Max Planck Society," explained Herwig. "Its members are representatives of the Federal Ministry of Research, the Bavarian government and so on, and in addition there were two or three independent scientists, including me. I was invited in the 1980s, when I was DG at CERN, not because I'm a plasma physics specialist, but because they have similar issues to large facilities like CERN." By this time, Schlüter had retired, and the director of the institute was Klaus Pinkau. "He was a very well-known physicist in Germany," said Herwig, "and someone I also knew from a long time before. When he retired, he was followed by Alex Bradshaw, a British man, who later followed me as president of the European Physical Society. The last director in the period I served on the Kuratorium was Friedrich Wagner. I had close contacts with all three of them and working with them was very enriching for me."

The IPP's main facility when Herwig was on the Kuratorium was ASDEX, the axially symmetric diverter experiment, a tokamak. "In a tokamak the plasma is enclosed in magnetic fields, but to heat it up you need a fast-changing field to heat the plasma by induction. This is done in a pulsed mode, tokamaks are pulsed machines, so long pulses are important. Another major research effort concentrates on controlling instabilities. Over the years the IPP has made fundamental contributions to these problems, and today is contributing to studies that feed into the big ITER project under construction in southern France, which is a world-wide collaboration including the USA and Japan."

During Herwig's time as a member of the Kuratorium, Germany underwent significant change with the fall of the Berlin wall and German reunification starting in 1990. This had repercussions across the country, and the IPP had an important role to play. "There was a push to boost scientific and technological activities in the new

Länder of the Federal Republic," said Herwig, "so the idea of creating a new institute there arose." The town of Greifswald, close to the Baltic Sea and gateway to Germany's two largest islands, Rügen and Usedom, with landscapes made familiar through the work of nineteenth century romantic painter Caspar David Friedrich, was chosen to host a new branch of the IPP. "The main problem was to find qualified personnel," recalled Herwig. "The only way to start that was to transfer personnel from Garching to Greifswald, but there was a lot of reluctance because everyone had families in Munich, and after unification infrastructure in the new *Länder* was not the same as in Munich." Nevertheless, an institute was established in Greifswald in 1994. "One key factor was that Friedrich Wagner volunteered to go and become the first director of the new institute at Greifswald."

The plans for Greifswald were exciting and ambitious. Tokamaks had been the main focus of magnetic confinement fusion since the 1950s, but time had shown that their pulsed operation was a major hurdle to viable energy production. A different kind of confinement device known as a stellarator, on the other hand, has a geometry that allows continuous plasma operation, but stellarators are not without their own challenges. The IPP decided to address them in Greifswald by building the world's largest stellarator. "Stellarators, as their name suggests, try to copy what happens in stars, where energy is created continuously by fusion, but building a stellarator is technically very complicated," said Herwig, "it requires a very complicated magnetic field. Stellarator fields can be calculated with modern computer programs, but their practical realisation is extremely difficult." Nevertheless, the Greifswald team came up with a proposal, and the machine was built. "It is called Wendelstein after a mountain near Munich," smiled Herwig, "that showed a certain nostalgia for the Munich laboratory among people from Bavaria, which is very mountainous, whereas Greifswald is completely flat."

Wendelstein 7-X, to give it its full name, produced its first plasma on 10 December 2015, and a vigorous programme of research has been underway since then. There has been huge progress in fusion research with tokamaks, and more recently, many promising results from Wendelstein 7-X, but viable fusion energy is still some way in the future. "Nevertheless," said Herwig, "fusion is a very attractive possibility for new energies. It does not involve uranium or plutonium, just hydrogen isotopes as fuel and helium as ash."

The European Atomic Energy Community, Euratom, had been interested in fusion research from the moment it was established in 1958. "In the framework of this programme, I was asked to join an advisory committee in Brussels in the early 1980s," said Herwig. "We had long discussions. Some of the industrialists said that the tokamak idea is bad because a public facility providing energy must be continuous, not pulsed, because that's not reliable. But then some engineers said it could be made reliable. I remember Heinz Riesenhuber, who was Minister of Research and Technology in Germany, and one day he asked me, "what are the outcomes of this committee in Brussels? How long do they think it will take to produce fusion energy in an industrial way?" I told him, "It's not so clear, but it seems certain it will take at least 30 years before we have an industrial kind of facility." "My God," Riesenhuber answered, "what a disaster, because yesterday I had to give a report to our parliament in Germany, and I told them it would be done within 20 years." So, that shows how difficult the communication between science and politics is.

Fig. 8.3 The Max Planck Institute for plasma physics in Greifswald, northern Germany, is host to the world's largest stellarator fusion device, Wendelstein X-7. The name is taken from a Bavarian mountain close to the original MPI for plasma physics in Garching, near Munich. The roofline is inspired by the waves of the nearby Baltic Sea (Courtesy of the Max-Planck-Institut für Plasmaphysik, photo: Ben Peters, ©Max-Planck-Institut, All rights reserved)

Today's big project in tokamak-based magnetic confinement fusion is ITER, the International Thermonuclear Experimental Reactor, at Cadarache in the south of France. It is scheduled to produce its first plasma in 2025. It should be the first tokamak to deliver more energy than is used to heat the plasma, a major milestone, but it will not deliver economically viable electricity to the grid. That is planned for the successor to ITER, DEMO, whose design will be shaped by results from ITER, and which could be delivering clean, economically viable energy by 2050. "Industrial fusion energy is still 20–30 years away," said Herwig, "but I'm confident that we'll get there, and when we do, particle physics will deserve some of the credit. For instance, the huge magnets they use need superconductivity, and the development of superconducting magnets was done to a large extent at CERN and other particle physics labs."

A Cuban Interlude

One project that Herwig worked on with UNESCO, although ultimately unsuccessful, served almost as a template for one of the greatest achievements of the third phase of his long career. It all started around 1996 in Cuba.

Russia, or more specifically, the Joint Institute for Nuclear Research (JINR) in Dubna, an international organisation, had made the gift of a small particle accelerator and it was being unpacked at the University of Havana. "They soon realised that they were not able to get the machine working without outside help," said Herwig, "although they had a quite good technical team." The idea of turning it into an international project serving the countries of central America soon emerged, and Cuba asked UNESCO to help. In turn, UNESCO asked Herwig to look at the proposal.

"Investigations in central American countries showed that there was indeed some interest in such a project," explained Herwig, "particularly in Mexico, so UNESCO decided to set up a special committee, and I was asked to chair it." Things got off to a good start, with meetings in Yucatan, the Cozumel Islands and in Havana. "At the beginning, it seemed quite possible to establish an international laboratory in Cuba." Things started to go wrong, however, when it came to finding a budget for the building that would house the accelerator. The budget required was around $600,000. "In Cuba there was a split budget," remembered Herwig, "separate parts for Cuban and US currencies, and buildings had to be paid for in dollars." The University of Havana did not have a dollar budget sufficient for this project and things did not look any better at the government level. "In discussions with the relevant ministers and other functionaries, I soon found out that they were unable to spare that many dollars."

None of the other central American countries involved in the project was prepared to put up the money, so a bid was made to the International Atomic Energy Agency (IAEA) in Vienna. "After very long discussions, this request was refused, and because Cuba couldn't find $600,000 and UNESCO did not have the budget either, the project

Fig. 8.4 In the mid-1990s, the JINR made the gift of a small particle accelerator to the University of Havana, and Herwig found himself charged with establishing an international laboratory around it. Despite his best efforts, funding was not forthcoming, and the gift was never put to use (©Anton Zelenov, Wikimedia Commons, CC BY-SA 3.0)

failed. As far as I know that accelerator is still sitting there gathering dust at the University of Havana," rued Herwig, many years later.

Herwig's Cuban experience nevertheless served as a template for things to come. Later, he and Maciej Nałęcz would turn their attention to a fledgling project to replicate the CERN model in the Middle East. A project that would go on to be one of the great successes of science diplomacy: SESAME (see Chap. 11).

IUPAP Looks into the Role of Women in Physics

In 1999, the American Physical Society (APS) celebrated its 100th birthday with a conference in Atlanta, Georgia with over 11,000 participants. These included members of the International Union of Pure and Applied Physics (IUPAP), which had chosen to hold its general assembly in the margins of the conference. The APS

Fig. 8.5 Herwig worked with many illustrious women physicists over his long career, including Lise Meitner and Chien Siung Wu. It is therefore perhaps not surprising that he was a member of the original IUPAP working group five on women in physics, photographed here during a visit to the Whitehouse in 2000 (©Holly Gwin, All rights reserved)

had also invited Cecilia Jarlskog who chaired the Nobel Committee for Physics at the time. At the IUPAP general assembly, she was seated close to the head of the Swedish delegation, Anders Bárány, and when the call came for any other business, she whispered in his ear that they should do something for women in physics. He made the suggestion, and that's how the IUPAP's Working Group 5 on Women in Physics was founded. Herwig was invited to join. "I was the only man among about a dozen women," recalled Herwig, "and I noted how different these meetings were compared to those that take place in all-male committees—much smoother discussions, though the outcomes were probably the same because whatever their debating style, scientists are first and foremost scientists with the same rational, evidence-based, approach to the world. All participants were against introducing quotas for women but agreed that their careers should be based on their achievements."

Forum Engelberg

Another organisation that sought out Herwig's help when he retired went by the name of Forum Engelberg. "At a time when science and technology were becoming ever more present in people's lives, someone like me who knows how science works and who also has experience in the management of smaller or larger organisations turned out to be much in demand," he recalled. "I believe that such a need exists today even more than it did 30 years ago."

"A former member of the Swiss delegation to CERN, Bernard Ecoffey, had the idea of creating a scientific forum with the aim of organising meetings to discuss general problems of global importance concerning technologies and society and to foster better relations between governments, industry and science," explained Herwig, "He asked me to chair it's scientific committee." Ecoffey chose a beautiful alpine resort to hold these meetings, and he named his initiative after it: Forum Engelberg. "Ecoffey managed to get funding from the Swiss federal and local governments as well as from industry, individuals and the EU," remembered Herwig. "I asked the French science minister, Hubert Curien, to be president of the forum, and we managed to attract some great speakers."

After about ten years in the role, Herwig stepped down, but he kept some precious memories from the meetings in Engelberg, and in particular, its historic Benedictine monastery. "A highlight of each forum was a dinner in the private rooms of the Abbot, Berchtold Müller, a very cultured man," said Herwig. "Over the years, I got to know him rather well, and he granted me a special favour. Occasionally he offered me a guest room in the monastery, from which it was just a few strides to the great church and its famous organ which had been played by Felix Mendelssohn-Bartholdy when he travelled on foot through Switzerland. I was given the key and at 5 a.m. when the church was not in use I was allowed to play. It was a miserable attempt since without any instructions I was unable to master the organ's several manuals and hundreds of

Fig. 8.6 The famous organ at Engelberg's Benedictine monastery, played by Felix Mendelssohn-Bartholdy during his travels through Switzerland, and by Herwig Schopper through his association with the Forum Engelberg (©W. Bulach, Wikimedia Commons, CC BY-SA 4.0)

registers. Forum Engelberg no longer exists, but the spirit of it continues. Today the Academia Engelberg [5] organises some very interesting dialogues."

The Foundation of the Cyprus Institute

One of Herwig's proudest achievements in retirement was his contribution to the foundation of the Cyprus Institute. "It all started on the initiative of just a few people," said Herwig. "I met Costas Papanicolas from the University of Athens at CERN in 1999, and he asked whether I would be prepared to participate in the foundation of a new university in Cyprus, a strongly developing country, but in great political difficulties because of its division into Greek and Turkish parts." Herwig agreed and at the beginning of 2000 took part in a meeting of a small group of people in a London hotel. "There were two representatives of the Cyprus Development Bank, John Joannides and Andreas Mouskos, who were prepared to provide some initial funds," recalled Herwig, "and the rest were academics: Costas Papanicolas, of course, Guy Ourisson who was president of the French Academy of Sciences, Frank Rhodes, who had served as president of Cornell University, Ernest J. Moniz of MIT, who later became US Secretary of Energy, and me. Papanicolas wanted the new university to be organised around departments covering different areas of research like MIT, rather than faculties, which explained the mixture of Europeans and Americans."

In that London hotel, this small group of people gave themselves the task of setting up the university by 2005. "During this planning period, we had many discussions involving the Cyprus government," said Herwig, "and it was decided that the Cyprus Institute should be a non-profit research and education institution with two complementary academic structures: a college, and the research centres devoted to special areas of technology." The institute operates under the aegis of the Cyprus Research and Educational Foundation (CREF), which is governed by a board of trustees and was established in 2004. One year later, the Cyprus Institute itself was formally established. "All the members of the founding team became trustees," said Herwig, "but unfortunately only three of us, Papanicolas, Moniz and me, lived to see the institute in full swing from 2007." Herwig was appointed as the first chair of a scientific advisory council and remains a member of CREF as a trustee emeritus. "The chair of CREF was taken for a number of years by Edouard Brézin, a physicist and former president of the French Academy of Science and is now occupied by Dan-Olof Riska, a professor at Helsinki University," said Herwig. "I proposed him for this post knowing him as chairman of the CERN pension fund. An executive committee is chaired by Andreas Pittas, who is very influential in the Cypriot economy."

Herwig, of course, made the most of his involvement with the Cyprus Institute, discovering the island, along with its history and culture. "Playing a part in the foundation of the institute was an extremely interesting experience," he explained. "Of course, the early years were rather busy since the institute was of great importance for the integration of Cyprus into the EU, so I had many occasions to visit the presidential palace in Nicosia. Over the years I got to know all the presidents and was impressed how much they believed in the importance of science regardless of their political colour. These meetings also gave me the opportunity to discuss and recommend the participation of Cyprus in CERN, and I'd like to think that I had some influence on the country's decision to join CERN."

Herwig's tourism in Cyprus carries bittersweet memories. "For me getting to know the Green Line between the Greek and Turkish parts of Cyprus was very emotional," he explained, "since it reminded me vividly of the Berlin Wall. My first crossing was somewhat complicated but over the years it has become easier. We hoped that the activity of the Cyprus Institute could be used to establish better relations between the two parts. I organised some private meetings with scientists from both sides of the line, but so far they have not really borne fruit."

"Costas Papanicolas was an essential guide to the history and culture of Cyprus," said Herwig. "Cyprus is almost unique, lying on the way between orient and occident, and offering sailors a place of rest. Over the centuries, many cultures have left their mark. I remember the surprise I felt when crossing from the Greek to the Turkish side to be confronted by a 13th century gothic cathedral built by the French Lusignan family, which is now used as a mosque. I also learned about how Richard the Lionheart conquered the island on his way to the Holy Land, apparently just to rescue his fiancée. Finally, I want to mention the beautiful landscape of Cyprus ranging from the high Troodos mountains to the Mediterranean shores, which I enjoyed on many occasions

Fig. 8.7 Pioneers of the Cyprus Institute at a meeting of the Institute's Board of Trustees. Left to right: Herwig Schopper, Costas Papanicolas, Director of the Institute, Andreas Pittas, Chair of the Executive Committtee and George Vassiliou, Former President of Cyprus and a founding member of the Institute (Herwig Schopper's personal collection. ©Herwig Schopper, All rights reserved)

on vacation with my wife. We developed very warm relations with my colleagues in Cyprus, and I can only hope that a solution will be found to bring the two parts of the island to live together peacefully."

Fig. 8.8 President of Cyprus, Demetris Christofi presented the Grand Cross of the Order of Merit of the Republic of Cyprus to Herwig at a ceremony in the Presidential palace in Nicosia in December 2012 (© Cyprus Institute)

In His Own Words: Exploring Cuba

"Working for the Central America project gave me a very interesting opportunity to learn about Cuba, and I got to know the country much better than I would have done as a tourist. I came to like it very much and went back several times on holiday with my family.

The first thing that surprised me when we arrived in Havana is that everywhere in hotels, and in offices, there were no pictures of Fidel Castro. After all, he was the Head of State, and it's normal in official buildings to see pictures of the Head of State. That was not the case in Cuba. I didn't see a single picture of Fidel Castro, but there were a lot of pictures of Che Guevara. I enquired why that was, and I got a lesson in Cuban history. I learnt that at the end of the First World War, US marines had been stationed on the island because the US invoked the Platt Amendment, signed in 1901. After the Spanish–American war of the previous century, Spain had ceded Cuba to the US, and although Cuba became independent in 1902, the Platt Amendment allowed the United States to intervene in Cuba. I learned that because of the Platt Amendment, although Cuba was independent, Americans exercised a lot of influence, and owned a lot of property. People told me that this sowed the seeds for the US-backed dictator Batista, and eventually the Cuban revolution that brought Castro to power in 1959. Cubans are not really enthusiastic communists. They never liked the influence of the Soviet Union very much, but on the other hand, they were really afraid of the US, fearing that another Batista could take over again. Maybe that's why they put up pictures of the romanticised heroic freedom fighter, rather than their Head of State? Of course, the revolution led to sanctions from the US, driving Cuba into the arms of the Soviet Union and giving rise to the economic difficulties the country still experiences today. One strange consequence I discovered is that you could go to any public telephone in the street, and call anywhere you wanted, but not the United States.

I was impressed by the charm of the country, which was flourishing when it first became independent. Many buildings in Havana reminded us of Paris, like the main avenue with hotels and all kinds of beautiful old palaces. We went to the places where famous visitors like Hemingway spent many years, taking our aperitifs in the same bars that he did. In Havana, I usually stayed in the Hotel Nacional, which is quite modern and close to the sea. The streets were full of vintage cars, hardly any new ones. For vacations, we usually went in winter, when Europe is freezing and Cuba has beautiful sunshine. We stayed in Varadero, a resort about 100 km from Havana and easily reachable by taxi where we could swim and sail. Another consequence of sanctions is that people would ask you for soap when you came out of your hotel.

On one occasion I paid a visit to Santiago de Cuba, which is an old town at the eastern end of the island that you had to get to by air. That was a real adventure. The plane was an old turbo prop, and besides the pilot, there always had to be a mechanic with all his tools so he could repair the plane if it got into trouble.

So, that's the background for my activities in Cuba. In conclusion, I'm still sorry that the UNESCO project to establish an international laboratory in Cuba failed, but my main regret is that very short-sighted western policy opened the door to Soviet influence in a part of the world where there was none. And I'm afraid that the present situation in the country still reflects this."

References

1. http://www.uniroma2.it/unesco/roste/eng.html
2. https://uia.org/s/or/en/1100023082
3. A world of science, vol 7(2). https://unesdoc.unesco.org/ark:/48223/pf0000181351.locale=en
4. https://www.euro-fusion.org/fusion/history-of-fusion/
5. https://academia-engelberg.ch/en/

Chapter 9
Travels to the Far East

China

When Herwig received his phone call from the office of Deng Xiaoping in August 1977, little did he know that it would lead to a long relationship with the People's Republic of China, and the Chinese physics community. "At that time, most people in the West didn't know much about China," recalled Herwig. "The cultural revolution had recently ended, and Deng Xiaoping was gradually assuming power in China."

Soon after, Herwig was invited to visit China, and he travelled there with his wife and son in spring 1978. "We flew to Beijing and were put up in the best hotel at the time, the Beijing Hotel, which was very close to the Forbidden City, and met, more or less, European standards." The purpose of the trip was to visit the Institute of High Energy Physics (IHEP), which had been founded in 1973 and was operated by the Chinese Academy of Sciences.

With the country emerging from the Cultural Revolution, the institute had the opportunity to develop. "I was warmly received, and I already knew some of the people from DESY," said Herwig. "They already had a linear accelerator, and they were discussing what they should do in the future. One of the possibilities they were considering was to build an electron–positron storage ring like DORIS, which we were operating at DESY at that time. That would be a very interesting machine after the discovery of the J/Psi particle because it would be able to study the so-called charm particles."

The Institute of High Energy Physics (IHEP)

IHEP had been established in Mao Zedong's lifetime, but with the Chairman's health failing in the early 1970s, a power struggle was already underway. By the time Herwig arrived in Beijing, it was close to being resolved in favour of the reformers, led by

© The Author(s) 2024
H. Schopper and J. Gillies, *Herwig Schopper*, Springer Biographies,
https://doi.org/10.1007/978-3-031-51042-7_9

Deng Xiaoping. By the time the Beijing electron–positron collider (BEPC) came on stream, making China's science world class was government priority. This was underlined by Deng's attendance at the ground-breaking ceremony in 1984, and again as the machine neared completion in 1988.

BEPC had a beam energy of 1.5–2.8 GeV, allowing it to be not only a factory for charm particles, but also for the tau lepton, which had been discovered in the mid-1970s. BEPC brought China on to the world stage in particle physics, providing, among other things, a precise measurement of the tau's mass. BEPC operated until 2004 and has since been succeeded by a more powerful machine, BEPC II, which provides luminosity around 100 times higher than its predecessor. "In recent years, they found possible hints of what is called violation of lepton universality, which implies different behaviour of different leptons, and goes against the standard model," explained Herwig. "This is a very exciting result, since there are very few experimental routes to physics beyond the standard model."

A Friendship Across Cultures

"One of the things that impressed me in China was their keenness to learn," recalled Herwig. "Everywhere I went they wanted to learn, learn, learn, and as we know now, they have done so with some success." Herwig had meetings with many people, and not only particle physicists. "I got to know the people at IHEP quite well, and in particular Chen Hesheng, who had spent time at DESY and became director of IHEP in 1998," said Herwig. "But one of the most important meetings of my life was not with a particle physicist, but with a politician called Fang Yi, a vice-premier who was responsible for science and technology. When we first met, it was not clear to me who he really was, but over the years I learned that he was one of the most powerful politicians in China. I think he played a great role in bringing China up to modern standards." Among his many roles in the Chinese state, Fang Yi was appointed vice president of the Chinese Academy of Sciences in 1976 and was close to Deng Xiaoping. From 1979 to 1981, he served as the academy's president. "He was not a scientist himself, but he understood very well the spirit and mentality of science," recalled Herwig. "One could talk to him as a scientific colleague."

Herwig first met Fang Yi in the Great Hall of the People on Tiananmen Square. "There were about 50 people placed in a u shape, all sitting in comfortable chairs with Fang and myself in the middle and an interpreter sitting directly behind us," said Herwig. "Only we two were allowed to speak, the others could only listen. There was no real discussion." Following that first official meeting came many more at which they discussed possible collaboration between China and DESY. "Fang was crucial in establishing relations between the People's Republic and the west," said Herwig, "and I think it's partly thanks to him that China developed so fast over the decades to keep up with the west and adjust to western technologies."

Herwig remembers Fang Yi as being a kind, easy-going and modest man, with even a touch of naivety. "When I told him that CERN was planning the SPS, he

Fig. 9.1 A photo taken during a visit by Herwig (front row, 4th from left) to China in 1977 to arrange collaboration between the Chinese Institute for High Energy Physics and DESY. The photograph was taken at a reception given by Fang Yi (front row 5th from left), who at the time was Vice-Chairman of the Academia Sinica. A powerful figure in China, Fang Yi went on to become one of the country's vice-premiers under Deng Xiaoping (©CERN archive, All rights reserved)

immediately said: 'Oh, we must also build one in China!'" Herwig persuaded him that it might be better to start with something more modest while China built up the necessary experience and expertise, and indeed that is what happened with the Beijing electron–positron collider, BEPC. "But times have changed now, and they are completely up to the west," said Herwig. "One possibility for the future of CERN is the future circular collider project (FCC), and the only real competition now is China. I believe what happens in Geneva at CERN will depend to a certain extent on what China decides to do."

During his many visits to China Herwig Schopper and Fang Yi reached a strong level of trust and mutual understanding. "To a certain extent," said Herwig, "I would say we even became friends in as much as it's possible for two people with completely different histories, traditions and mentalities to be friends. One important factor is that he spoke reasonably good English, so we did not need an interpreter. If you can only talk through interpreters, you never really establish a personal contact. We stayed in contact for many years, even after I left my job as Director-General at CERN. He became ill and we exchanged letters and little presents while he was in hospital. I'm sorry to say that he died in 1997 at the age of 81."

Bismuth Germanium Oxide for the L3 Experiment at CERN

While serving as Director-General of CERN, Herwig's travels in China took him to Shanghai where he visited the Chinese Academy of Sciences' Shanghai Institute of Ceramics. At the time, Sam Ting was leading the team responsible for the L3 experiment being prepared for CERN's Large Electron-Positron collider (LEP), which was planning to use a novel scintillating crystal approach for its electromagnetic calorimeter. The Shanghai Institute was world leading in producing such crystals. "I visited several important laboratories in Shanghai," recalled Herwig, "including one where they were able to fabricate so-called BGO crystals. BGO stands for bismuth germanium oxide, sometimes also called bismuth germanate."

Scintillators emit photons when struck by ionising particles, and this can be used to measure the energy of those particles. BGO is particularly good in this respect, and Sam Ting's L3 experiment needed about 20,000 large crystals for its electromagnetic calorimeter. "In order to work as well as L3 required, the crystals had to be very pure," explained Herwig. "They had to be made in cleanrooms, but even the best cleanrooms in the west, largely developed for the production of electronic chips, were not good enough to make BGO crystals. It turned out that the Chinese were the only ones who could make BGO crystals that fulfilled all the specifications of Sam Ting's detector."

It was curiosity that took Herwig to Shanghai: he wanted to visit the institute that could make cleanrooms better than any that the western electronics industry was capable of. "I was surprised," he remembered. "I was taken to a simple wooden barrack that they had converted into a cleanroom. It was impressive how the Chinese could achieve things with very basic means that modern technology was incapable of, simply by being very careful." L3 was a pioneer in the use of scintillating crystal calorimetry, which is now a widely deployed technique. The CMS experiment at CERN's LHC, for example, uses close to 80,000 lead tungstate crystals in its electromagnetic calorimeter.

Herwig's trip to Shanghai was memorable as much for what he did not expect to find there as for what he did. As its name suggests, the Shanghai Institute of Ceramics carries out research into that most iconic of ancient Chinese inventions, porcelain. "They really understood the secrets of ancient Chinese porcelain manufacture, and they knew how they could produce new porcelain the same way as it was done hundreds of years ago," said Herwig. "They had also developed methods to find whether a piece of porcelain is genuinely old or a modern fake copy. For instance, old porcelain reacts differently to illumination by x-rays or ultraviolet light than new porcelain." Herwig also learned that the institute could produce porcelain that would fool the experts. "So I asked what they did with such products, and they told me that they labelled them and sold them to Hong Kong, adding with a smile that they didn't know what happened to the label after. I decided never to buy old Chinese porcelain."

A Memorial to Chien-Shiung Wu

Herwig's last trip to China was in the early 2000s when he was invited to visit Nanjing, where Chien-Shiung Wu had gone to university. Because Herwig himself had performed pioneering early experiments on parity violation, he felt a strong affinity with the first lady of physics, as Madame Wu had become known. "Because we had each carried out an experiment proposed by Lee and Yang, we met very often afterwards and became friends," he explained. "I also got to know her husband, a Chinese-American physicist called Luke Chia-Liu Yuan, and we met very often privately later."

Chien-Shiung Wu died following a stroke at the age of 84 in New York city on 16 February 1997, but she chose to be buried in her home village of Liuhe on the Yangtse River between Nanjing and Shanghai, where Luke was to join her when he passed away in 2003. Today, a memorial marks the place where the ashes of Chien-Shiung Wu are buried, in the courtyard of the school, founded by her father, that she had attended as a small girl. After her death, her alma mater, Southeast University in Nanjing constructed the Chien-Shiung Wu memorial hall [1], which contains an exhibition based on donations from her husband and from Columbia University. "There was a lot of discussion about why she didn't get the Nobel Prize," said Herwig. "She should have received it, but she got many other honours during her life, among them the highest American honours that can be given to a scientist, and this museum and meeting centre in her name was to be the last. I felt deeply honoured to be invited to speak at the inauguration."

"This was my last trip to China, and on this occasion, I got to know the modern China," concluded Herwig. "Of course, in many respects, social conditions have improved enormously, but it has also become very western. Airports are like everywhere in the world. The streets are full of cars like everywhere in the world. So, I find that the old Chinese charm has gone. I developed a certain admiration for the old culture, and for the way they adapted to the modern world. And I don't have to tell you that in future, China will play a major role in world history."

Taiwan ROC

In parallel with his visits to the People's Republic of China, Herwig also had several opportunities to visit the Republic of China on the other side of the Taiwan Strait. In 1949, following the defeat of Chiang Kai-shek's nationalist Kuomintang armies by Mao's communists in mainland China, the nationalists took refuge on the island of Taiwan and groups of smaller islands close to the Chinese mainland. Taiwan became the refuge of several million Chinese partisans of the Kuomintang, and Taiwan thus became the seat of General Chiang's nationalist republic. The Kuomintang ruled as a single-party state for 40 years, with democratic reforms only taking root in the 1980s. The timing of Herwig's visits to the island were therefore timed to give him an ideal vantage point to witness another society in transformation. "I paid a few visits to Taiwan, which allowed me to follow the fantastic and breath-taking rise of this country from a completely underdeveloped state to one with a most modern and competitive economy," he recalled. "My first visit was in the 1970s when I was at DESY. The Taiwanese wanted to develop their science and considered building either an electron–positron storage ring, like DORIS, which we had at DESY, or a synchrotron radiation source." There was no doubt that the young republic had the intellectual firepower to do so: when Chiang Kai-shek fled to Taiwan, many scientists, engineers and other intellectuals joined him. "They were crucial to transforming an agrarian economy into a modern industrialised society over the following decades."

At the time of Herwig's first visit, Taiwan had started to build a small linear accelerator that could be used as an injector for a bigger machine. "They invited me in order to discuss a scenario that would work for them," said Herwig. "I advised them to build a synchrotron light source which would have a broader field of application, and later I was impressed by their determination, their engagement, and their ability to learn quickly and put plans into action."

Chiang Kai-shek died in 1975 and was succeeded by Yen Chia-kan, but it was General Chiang's son Chiang Ching-kuo who was the de facto leader. He became president in 1978 when Yen resigned and remained in office until 1988. He set the tone for the liberalisation of the republic through the 1990s, increasing freedom of speech and tolerating political dissent, but he also had to deal with an evolving international political situation that seemed to favour the mainland. In 1971, for instance, China's seat at the United Nations shifted from Taipei to Beijing, leaving the republic in an ambiguous and tense diplomatic situation that prevails to this day.

"This was the political background when we built LEP at CERN in the 1980s," said Herwig, going on to reintroduce a regular protagonist in his career. "Sam Ting suggested that both the People's Republic and the Republic should participate in the L3 collaboration. In spite of the very critical political situation this was approved by both governments at the highest level. It was the first time that scientists from the PRC and the ROC could work together in the same experiment, and a remarkable example of science for peace."

As a result of Ting's initiative, Herwig visited Taipei in his capacity as CERN's Director-General. After the official business was over, he was treated as a privileged guest of the authorities. "I remember a visit to the National Palace Museum in Taipei," said Herwig. "When Chiang Kai-shek left Beijing after the war, he took with him many cultural treasures from the forbidden city and they are now on display at this museum, which is probably one of the largest and most precious collections of Chinese arts covering several thousand years. The most precious objects of the many thousand displayed are protected in heavily guarded vaults."

As on the mainland, Herwig's admiration of the Chinese culture he found in Taiwan was accompanied by a strong dose of culture shock. "One evening I was invited by a minister to an official dinner consisting of some 20 courses," he recalled. "We arrived at course 19 and I had not eaten a single bite. Taiwan is full of exotic dishes, which I struggled with, but I was saved by Sam Ting, who ordered a special serving of rice so I could eat. Noticing my discomfort, the minister lightened the

Fig. 9.2 Towards the end of his mandate as CERN Director-General, Herwig had discussions in Taipei with Lee Teng-hui (centre), President of Taiwan, on improving scientific cooperation between Taiwan and Europe. Right is Chen Li-an, President of the Taiwan Commission for Sciences (©CERN archive, All rights reserved)

mood by telling a story of his experience of discovering a new culture in America. Once he came to New York and wanted to buy a present for his wife in one of the famous jewellery shops. When he tried to pay by credit card they refused since he was not a known client. When he was about to leave the shop without a present the employee called him back and told him that he had noticed that the minister was wearing a Rolex watch and this was considered as sufficient evidence for his financial credibility. This amused the minister, who told us that his watch was a fake bought in Hong Kong for 5 dollars."

Japan

DESY and CERN had many connections with Japanese laboratories. "I made many visits during my terms of office at DESY and CERN," Herwig said, "in particular to the KEK laboratory and the University of Tokyo, which is considered as one of the elite universities in the world, and I made one particularly good friend in Toshi— Masatoshi Koshiba—who won the Nobel Prize in 2002 for his pioneering work on neutrinos. Earlier, he spent a year as a visiting professor at DESY and our families became close friends. He was about my age, and I was greatly saddened to learn of his passing in 2020. Toshi came from an old Samurai family, a kind of nobility in Japan, and was in some respects very traditional, in others quite liberal. His apartment in Tokyo was quite modest as property in Japan is scarce and expensive. Once he invited me with my colleague, Erich Lohrmann from DESY, and our wives to a

Geisha evening. Much misunderstood in the west, I learned that traditional Geishas are highly respected in Japanese culture with an extraordinary education in the arts, literature and music. Their main task is to entertain guests by singing, conversation and dance, and above all by exhibiting a traditional costume worth a fortune. On another occasion, he invited us to an apartment by the sea with the great luxury of having a natural hot spring in the house, but again everything was tiny. It had everything you needed, but nothing to excess. He also introduced me to the famous Japanese tea ceremony. He was a wonderful guide to Japanese traditions and culture. Later Toshi spent some time with his family at CERN in Geneva, and my family tried to reciprocate. In the meantime, his daughter had grown up and was supposed to marry. According to traditional Japanese customs the future husband is selected by the parents. However, Toshi was sufficiently liberal to accord to his daughter the right of veto. After she rejected the first and second choices, she was again presented with the first, and this time she said yes. Really, I think she had a say from the start, and as far as I know, it became a quite happy marriage."

Fig. 9.3 The 2002 Nobel Laureate, Masatoshi Koshiba, addresses an eminent audience in a packed CERN Auditorium on 8 July 2003 (©CERN, All rights reserved)

The Subcontinent: Pakistan

Another occasional destination for Herwig, before and after retirement, was Pakistan. "The country has a rather well developed scientific and technical environment," he explained, "which is probably due to some extent to the British occupation leaving behind a relatively good education and administrative system, so it's not surprising that Pakistan started cooperation with CERN and also played an essential role in the development of SESAME" (see Chap. 12).

Pakistan signed a cooperation agreement with CERN in 1994 and became an associate member state of the organisation in 2015. In between, the country made notable contributions to the CERN programme, and developed plans to establish a synchrotron light source in the country. In 2006, Pakistan's controversial president, Pervez Musharraf, visited CERN, and Herwig came to know first-hand his ambitions for science in Pakistan. "Pervez Musharraf was considered as a sort of dictator, having come to power by military means, but he was a secular president, he was interested in establishing good relations with the west, and he promoted science and technology," explained Herwig. "He invited me to Pakistan several times and made it possible for me to see parts of the country that are not open to tourists. With him and other government representatives we had many discussions about the long-term scientific development of Pakistan."

Fig. 9.4 Pakistan's President, Pervez Musharraf, visited CERN on 25 January 2006. He is seen here signing the Laboratory's guest book, the *Livre d'Or*. Herwig came to know of the scientific ambitions Musharraf had for his country and had the chance to visit on several occasions (©CERN, All rights reserved)

Herwig met Musharraf on several occasions in Islamabad where they discussed cooperation with CERN and the possibility of constructing a synchrotron light source in Pakistan. By this time, Herwig was already heavily involved with the project to build a light source for the Middle East and neighbouring regions, a project known as SESAME, or synchrotron light for science and applications in the Middle East, as an intergovernmental organisation on the CERN model. Herwig argued that Pakistan should join SESAME before building its own light source.

Scientific research in Pakistan falls under the remit of two organisations both based in the capital, Islamabad: the Pakistan Atomic Energy Commission (PAEK), and the Pakistan Science Foundation (PSF). Through them, Herwig was able to visit the production sites of several of Pakistan's contributions to CERN. "The supports of the ring magnets for CMS were produced at a machine shop located some distance away from Islamabad in a kind of desert," said Herwig by way of example, "I was impressed by the installations, but the area is also interesting from a historical point of view since Alexander the Great had reached this region and had left some traces." When travelling outside Islamabad, Herwig was always accompanied by a military vehicle. "At the time I visited Pakistan it was relatively safe, in particular Islamabad, where I could move around the centre completely freely at any time of day. I visited a very big modern mosque whose construction had been financed by Saudi Arabia. Of course, some places were to be avoided, such as the area above Peshawar, which was a centre for the Taliban in Afghanistan."

"I was invited twice to international conferences at Nathiagali, a beautiful resort in the mountains at a height of about 2500 metres in a fantastic mountain landscape," Herwig recollected. "It had been used by the British as a summer government retreat, and a modest meeting place had been established there, where I gave talks about CERN and SESAME."

India

India's involvement with CERN goes back to the 1960s, when scientists from the Tata Institute of Fundamental Research in Bombay, now Mumbai, first started to visit the European laboratory. "This was mainly due to the outstanding Indian theoretician Homi Bhabha," explained Herwig. "He became the founding director of the Tata Institute, which was established after World War II, since even before independence, India recognised the importance of natural sciences. The emphasis of the institute was theoretical physics and also some experiments to observe cosmic rays." Following independence, the Tata Institute became a reference point for intellectual excellence in India. "When I travelled around the country, I found that everywhere I went, educational establishments held up the Tata as the model to aspire to," said Herwig. "When I visited the institute for the first time, I was shown the former office of Bhabha which is kept in its original state, almost like a sanctuary. To my surprise I saw a copy of a preprint from DESY dealing with Bhabha scattering on his table."

Fig. 9.5 Homi Bhabha (second from right) in the Palais des Nations in Geneva for the opening of the first International Conference on Peaceful Uses of Atomic Energy on 8 August 1955. He is flanked by (left to right) Swiss President, Max Petitpierre, UN Secretary General, Dag Hammarskjold, and Conference Secretary General Walter G. Whitman (©IAEA, Wikimedia Commons, CC BY-SA 2.0).

Homi Bhabha was a passenger on Air India flight 101, which crashed into Mont Blanc en route from Bombay to London on 24 January 1966 with the loss of all on board. At the time, there was speculation that the crash was not an accident, because it was carrying the man who came to be known as the father of India's nuclear programme. The subsequent investigation concluded, however, that the plane had a faulty VHF receiver and was relying on verbal instructions. The pilot had started his descent for a scheduled stop in Geneva before clearing the mountain. Bhabha, it seems, along with 126 others on board, lost their lives because of a simple miscommunication.

Herwig never met the father of India's nuclear programme, but he remembers good relations between CERN and the institute for fundamental research that he created. "I met his successors, M. G. K. Menon, B. V. Sreekantan and Virendra Singh, who ran the institute from 1966 to 1997," said Herwig. "Once, I was part of a guided tour of Mumbai, which included a visit to the Tata. There, the guide pointed to a statue of Shiva showing the eternal dance of the elements of nature, and the guide explained that the dance of the elementary particles in the atom are also presented by this dance—a wonderful unification of natural science and religion." Today, a similar statue stands at CERN, a gift of the Indian government in 2004.

As in China, Herwig profited from his business trips to India to explore the country and its cultures. "I travelled around India on occasions, sometimes after conferences or just as a tourist. My wife and I saw beautiful places, like the Taj Mahal at Agra, and Madras. We admired Indian art and culture and would sometimes spend the night in luxurious former palaces of maharajas, or at the other extreme in the rather more modest accommodations of a Guru. I do not want to repeat here just what is described in travel books, but I can't help recalling the cultural dissonance of the colonial legacy, which was particularly striking in Goa, the former Portuguese colony. I attended several conferences there and was always struck by the presence of a beautiful gothic cathedral in the midst of an Indian city. It was the same when I visited Macao."

Both India and Pakistan carry out their fundamental physics research in part through institutes dedicated to atomic energy, and both are today nuclear powers. Born from the calamitous partition of British India in 1947, which displaced millions along religious lines, the two countries have always had a tense relationship. Their involvement with CERN led to Herwig being invited to take part in a meeting on the shores of Lake Como organised by the Rockefeller Foundation. "The main topic was the threat of a nuclear war between India and Pakistan," recalled Herwig. "Since by this time, both countries had nuclear weapons, there was great concern that they would use them. I brought in the argument that they wouldn't dare since doing so would lead to self-destruction. Both had developed such weapons for defence, and for national prestige. I remember introducing the probably unrealistic idea that nuclear war could be replaced by cyber warfare, which would carry less of an existential risk. Such an argument holds in principle for any country developing nuclear arms."

Vietnam

The last Asian country that Herwig developed links with is Vietnam. "Sometimes one gets involved by chance, and for me this was the case with Vietnam," said Herwig. "At CERN I got to know a French physicist originating from Vietnam. His name was Jean Trân Thanh Vân and in his activities he was strongly supported by his wife Le Kim Ngoc, also a physicist who has done much for Vietnam in her own right."

Jean Trân Thanh Vân was born in Vietnam in 1936 and moved to France in 1953, where his successful career led to him to becoming a director of research at the National Centre for Scientific Research (CNRS) in 1991. Despite his success in Europe, he never forgot the country of his birth.

Beyond research, his forte is as an organiser of meetings, establishing the Rencontres de Moriond in 1996, a conference series in the French Alps which has become one of the most important global conferences for particle physics. In 1989, he repeated the model for the field of astrophysics with the Rencontres de Blois, and when Vietnam started to open up to foreign scientists in 1993, he established the Rencontres du Vietnam, to provide a forum for dialogue between the global scientific community and Vietnam. From the Rencontres du Vietnam grew the beautiful International

Fig. 9.6 Herwig shakes the hand of Tran Dai Quang, President of Vietnam, on the occasion of the 12th rencontres du Vietnam in July 2016 (Herwig Schopper's personal collection. ©Herwig Schopper, All rights reserved)

Centre for Interdisciplinary Science and Education (ICISE), located where the mountains meet the sea at the town of Qui-Nhon, between Hanoi and Ho Chi Minh city. "Vân invited me to conferences at the centre and I participated twice," said Herwig.

Herwig spoke about CERN and SESAME as role models for successful international collaboration in science, and on his second visit, in spring 2018 raised the question of creating such a centre for Southeast Asia. "It was a proposal that was accepted with great interest," recalled Herwig. "Vân had developed excellent relations with the government. After each conference the participants were invited to a meeting at the presidential palace in Hanoi in the presence of government ministers and the president himself." These were very formal meetings at which, following the president's address, three guests were each given three minutes to speak about their ideas for Vietnam's development. "At the second meeting I attended, the speakers were two Nobel Prize winners and me. I used my three minutes to suggest creating an international research centre on the CERN model, like we had done with SESAME in the Middle East, promoting science and technology and at the same time smoothing relations between countries." There was no discussion, so Herwig received no immediate reaction to his proposal. "But after the meeting the minister of research approached me, asked some detailed questions and promised that they would follow up this idea" he recalled. "Later I learned that the discussions continued, and that Singapore was interested to participate or even house a research centre for synchrotron radiation."

"Nothing has so far come of this, but these things take time, and I hope that one day it will happen," said Herwig. "These visits gave me an excellent possibility to get to know the beauties of this country, in particular an excursion to the famous bay of Vinh Ha Long with its multitude of islands traversed by picturesque sailing junks. I noticed that in spite of frictions between Vietnam and the People's Republic, there were many Chinese tourists, and new hotels being built to accommodate them. Perhaps tourism can work for peace, as well as science?"

In His Own Words: Memories from a Big Country

"Beijing in the late 70s was a grey place." At that time, just after the Cultural Revolution, everyone was dressed in Mao suits, grey suits. There was no colour to be seen on the streets or anywhere. There were lots of people and the roads were crowded with bicycles. There were hardly any cars, just a very dense traffic of bicycles. There were so many that I was puzzled, and I asked, 'What happens if there are accidents between two bicycles? Do people have to pay a fine?' I was told, 'No. In China there are no fines. If somebody causes an accident, then he will be summoned to the team where he works. There will be a discussion about the error he committed, and in that way, he will learn not to make the same mistake again.

Things were quick to change, and when I came back ten or maybe 15 years later things were very different. There were practically no bicycles anymore, but only cars. So I asked again: 'What happens if there's an accident between two cars, are people still summoned to their work team to be educated?' I was told, 'No. Not at all. They have to pay a fine and that's it.' Things change.

Of course, in the '70s, there were practically no foreigners in Beijing, so my family and I were among very few, which made us quite exotic. While I was at work and in meetings, they invited my wife and my son to visit the zoo to see the pandas, which were famous around the world at that time. Everyone turned to look my wife. When she remarked that people seemed more interested in her than the pandas, the interpreter explained that they had never seen anyone with blonde hair and blue eyes before! On another occasion, my son was taken on a trip accompanied by a young Chinese person, and they were put up in a modern hotel for foreign tourists. It had a bar with neon lights, a band playing modern rock music and people drinking whisky. My son's companion was completely flabbergasted: he'd never seen anything like it. Things like this were among the first signs of western culture appearing in post-Cultural Revolution China.

Of course, during my various visits to Beijing, I visited many of the interesting places that all the tourists go to these days, like The Great Wall and the Ming tombs. They're so well known these days that I don't need to go into detail, but one of these visits says a lot about the Chinese mentality. Not far from Beijing are the famous Ming tombs, the mausoleums of 13 Ming dynasty emperors dating from as far back as the fifteenth century. In imperial times, ordinary people could not visit them, and even today, only a few are open to visitors. One reason, I was told, is that the Chinese archaeologists decided not to explore them all because opening up an old tomb is a one-off event, which cannot be repeated, and since technology for archaeology is always improving, it would be irresponsible to open them all now—far better to leave them to future generations. The wisdom of that impressed me very much. I can't see us in the west taking such a long view. Our curiosity would lead to us opening them all to see what's inside.

Walking around in Beijing was a bit of an adventure: everything was written in Chinese, and very few people spoke English. One time I got lost and had no idea how to get back to the hotel. Nobody on the street could understand me, but I could recognise pharmacies, so I went into one hoping that the pharmacist might speak some English. To my great relief, he did, and was able to explain to me how to get back to the hotel. On one of my trips, I bought a model of one of the famous grey horses from Xi'an as a souvenir. It was about the size of a big dog, which would be difficult to bring home today, but back then I was able to just carry it onto the plane under my arm. Those were happy times to travel by air—few controls or checks.

When I visited Xi'an and the tomb of the first emperor with its terracotta army, one nice thing was that there were practically no tourists, no foreign tourists anyway. Thanks to the authorisation of Fang Yi, I was able to visit not only the excavations, which were not as far advanced as they are today, partly because of that Chinese long-view I learned about at the Ming tombs, but also the laboratory where they investigated the objects they had excavated. They were very proud to show me a chariot, a metal chariot that they had just excavated, and which was not yet on display. Their analyses had shown that it was made of metal alloys that were not even known at the time.

I couldn't resist discussing with my Chinese colleagues the importance of traditional medicine compared to modern medicine. In particular I asked what they would do if they got seriously ill. Some of them told me: 'Well, of course you would go to a modern hospital. Chinese medicine only works if you believe in Mao.' They said it with a laugh, of course, but I think that was also a sign of the changes underway in China at the time. When they weren't being so flippant, we all had to agree that Chinese medicine is based on many centuries of practical experience, and worthy of being taken seriously.

On several occasions, I visited a place that nobody in the west had heard of, but we all know it now: Wuhan, and I saw the now infamous live animal markets where Covid-19 most likely made the jump to humans. I was there for conferences lasting one or two weeks, and the organisers took us for boat trips on the Yangtze River. The first time I went on one of these trips, the scenery was beautiful, but the river was squalid, and lined with coal-powered factories producing a lot of smoke. We were fed on the boat, and one of the soups contained frogs. We even saw a human cadaver floating in the river. Things were very different the next time I went. We had a modern boat, and everything was clean. We could enjoy the beautiful landscape where the river flows through the three gorges without distraction. Since then, they've built the famous three gorges dam to tame the water and use it for electricity production. When I was there, construction of the dam was in full swing. It was very impressive, and it's still probably one of the largest industrial projects ever made in China. When I was first there, they had just started it. The second time, they had started to fill the reservoir with water. Of course, the whole idea of the dam is controversial, but the people I met there were all in favour of it because before the dam it was normal that every decade or every two decades there would be a big flood, killing people and wiping out crops. Now, because of the dam, they can control the flooding. Of course, there are also negative effects on the environment, but I think the people who live there are quite happy to have the dam.

On one occasion, I was invited to visit an aunt of Samuel Ting. She was a widow, and her husband had been a participant on the Great March, so she belonged to a politically favoured family. This meant that she was allowed to live in a very privileged quarter of Beijing very close to the Forbidden City. The houses were very modest, humble even, but still it was a very privileged quarter. Today I think it has been completely replaced by modern buildings, but when I was there, I was surprised at how modestly even such a privileged family lived. The rooms were rather empty. There was not much furniture. Nothing comfortable on which one could rest, just a table, chairs and not much else. It was quite an occasion for Sam's aunt to have a western visitor, so she'd also invited her sisters, and there were six aunts in total. To be kind to me, they didn't introduce themselves by their Chinese names, but as aunt number 1, aunt number 2, and so on.

China is a vast and very populous country, we all know that, but one event really brought it home to me. When I was Director-General of CERN, I visited China and met a vice-president of a firm that produced non-ferrous metals like niobium, because we needed niobium for the radiofrequency cavities of LEP. Over dinner he asked me what CERN is, and how big it is. I explained it to him, and then I asked him how big his company was. 'We are a relatively small firm in China,' he said. So I asked him how many employees he had. Two million, came the answer.

I already had a feel for the sheer size of the country. At the end of my first visit, I was invited by Fang Yi for a trip from Beijing to Canton, as it was at the time, and ending up finally in Hong Kong. I came to understand that China is not just a country but a continent, a huge part of the world full of different ethnic groups speaking many languages and dialects. It's for that reason that the Chinese government insists that all children learn Mandarin because otherwise they would not have a common language. They also insist on a universal script. When we travelled through the country from one province to the other, sometimes interpreters had difficulty speaking with local people, and when that happened, they pointed to the written characters, and these of course were understood.

Everywhere we went, the local dignitaries wanted to meet me. They were hungry for knowledge, and they squeezed me for all the information I could give them. They took me to all kinds of factories where they asked me for advice on everything from how to make steel to building transformers. I had to disappoint them to a certain extent, but I did try to give them advice on managing an organisation.

Shanghai was particularly interesting. I went there for various conferences and was impressed, maybe even shocked by the extent of western influence. Because of the importance of Shanghai's harbour, and the international settlement that was run by America and Britain right up to the 1940s, western influence was everywhere. The waterfront known as the Bund could easily have been in America or the UK.

All in all, I have very strong memories of China, of the many people I met there, and I feel very privileged to have had the opportunity to spend time in the country at a time of such profound change."

Reference

1. https://www.seu.edu.cn/english/2016/0614/c238a161657/page.htm

Chapter 10
The Large Machines: LEP, the LHC and Beyond

Although no longer at CERN after his retirement, Herwig maintained a keen interest in the Large Electron-Positron collider (LEP) throughout the machine's operational lifetime. "I considered it as a kind of child," he explained, "and with one member of my real family staying in Geneva, Andreas by this time was working on the pioneering CPLEAR experiment studying CP violation at CERN, we kept our home in Switzerland."

Following the excitement of LEP's first big result, announced on 13 October 1989, which had shown that there are three and only three families of fundamental particles in nature, each of the four LEP experiments went on further to pin down that number to 2.984 plus or minus 0.008. Why there are only three families remains, however, a mystery and a question for future research to address. Other key results were soon to follow, building LEP's legacy as the machine that put the standard model of particle physics on firm experimental foundations.

LEP ran in two phases: LEP I until 1996, and the higher energy LEP II until the machine was finally switched off in 2000. Each phase was designed to play a specific role. LEP I was a Z factory, producing Z particles in great numbers so the experiments could pin down its properties and thereby put the underlying electroweak theory to the test. LEP II took the collision energy above that needed to produce charged W particles in pairs, so the experiments could put them to a similar test. "By 2000, thanks largely to LEP," said Herwig, "the electroweak theory was confirmed as being among the greatest intellectual achievements of twentieth century science. The carrier W and Z particles had been measured with precision, and all the theoretical predictions about how they should behave and interact with other particles put to the test. LEP had transformed particle physics from a 10% accuracy science to a precision science with errors smaller than 1%."

On the political front, there were important developments through the LEP era too. Early in his mandate as CERN's Director-General, Carlo Rubbia presented plans to the CERN council that foresaw a large hadron collider (LHC) installed in the LEP

© The Author(s) 2024
H. Schopper and J. Gillies, *Herwig Schopper*, Springer Biographies,
https://doi.org/10.1007/978-3-031-51042-7_10

tunnel and running along with LEP by 1998, but it was not to be. At least not in that form, or on that timescale.

Across the Atlantic, work began on the ambitious Superconducting Super Collider (SSC) project, a hadron collider over 80 km around with a projected collision energy of some 40 TeV. It too was not to be: Congress pulled the plug in 1993, leaving the high energy future for global particle physics looking bleak.

Throughout Rubbia's mandate, however, R&D continued for CERN's LHC, refining the design and reducing the costs. As a result, when Chris Llewellyn-Smith succeeded Rubbia as CERN's Director-General in 1994, the project was still very much alive. In December 1994, Llewellyn-Smith went to the CERN council with a plan for a reduced-cost LHC that would be built in two stages to spread the cost over a longer period of time, while simultaneously paring back the rest of the laboratory's research programme to a minimum. The plan worked, and the Council gave its blessing. The LHC was planned to start up in 2004 with only two-thirds of its magnets in place [1], and therefore able to operate at two-thirds of its design energy for a number of years before installing the remaining magnets and moving up to design energy. A decision on the precise schedule was deferred until 1997, but was eventually taken in 1996. By then, CERN had secured substantial support from non-member state countries, notably the USA, which, with the cancellation of the SSC was looking to Europe for its future, and Japan. Coupled with contributions from Russia, India and Canada, this global support for the project emboldened the CERN Council to approve construction in a single phase, albeit within a reduced overall budget for CERN. "This was a smart move by Chris," said Herwig, "very politically astute."

While all this was going on, LEP I concluded with around 18 million Z particles recorded and analysed [2]. By 1996, new superconducting cavities had been installed around the ring to boost the machine's energy to above the W-pair production threshold, with the collision energy gradually being increased over time towards its maximum of 209 GeV, a record for an electron machine to this day. By the year 2000, the LEP experiments had recorded some 80,000 W-pairs, and the experiment's job of precision testing the electroweak theory was reaching a conclusion. Civil engineering works for the LHC had been on-going in parallel. The time had come for LEP to make way for the LHC. But there was to be a twist in the tail.

A Nail-Biting Finish

By May 1999, a total of 288 superconducting accelerating cavities gave LEP the capacity to achieve a collision energy of 192 GeV. This was nominally the maximum energy that the machine would reach, with LEP due to be switched off at the end of the year, but circumstances conspired to change the course of events. First of all, although the mass of the Higgs boson was not predicted by theory, the range of masses available to it could be constrained by ever-more precise measurements of other parameters in the Standard Model. By 1999, such measurements had constrained the mass of the Higgs boson to be in the range of around 90 GeV to about 200 GeV, with the probability dropping with increasing mass. In other words, LEP had entered the most likely energy range to find this most elusive of particles.

In 1998, the pressure to keep LEP running was already high. Civil engineering work for the LHC was rescheduled in such a way as to give LEP one more year of running without delaying the LHC's start-up, and the CERN Council agreed to keep LEP for one more year. As 1999 drew to a close, ways were found to push LEP's superconducting cavities beyond their design limits. By November, they were delivering collisions at 202 GeV. 2000, LEP's extra year, was poised to be an exciting one.

The beams of LEP I had been accelerated by 128 normally conducting copper cavities, which gradually gave way to the superconducting ones as LEP I transformed into LEP II. At the end of 1999, 48 copper cavities remained in the ring, but there was plenty of room to reinstall eight more and push the collision energy up to 209 GeV from May 2000. With the Higgs appearing tantalisingly within reach, every extra GeV of energy mattered.

Already with 202 GeV collisions in the bag, two of the four LEP experiments were reporting potential candidates for Higgs bosons in their data. Regular meetings, each one packed to the rafters, were scheduled throughout the year for the experiments to report their latest analyses. "The Higgs candidate events remained, but the other two experiments continued to see nothing," remembered Herwig, "nevertheless, combining the results of all four experiments led the LEP experiments committee to conclude in November that the data were compatible with a Higgs particle with a mass of about 115 GeV, with a likelihood of around 50% that the measurement would stand the test of time." Those may sound like good odds, but in physics, signals such as that come and go with alarming frequency.

In July 2000, Fermilab had announced the discovery of the tau neutrino, leaving the Higgs boson as the last missing ingredient of the Standard Model to be discovered. With Fermilab's Tevatron scheduled to start its long-awaited Run II in 2001, with increased energy and luminosity, the stakes could not have been higher.

CERN's management, headed by Luciano Maiani since 1999, had a difficult decision to make. If they decided to run LEP for another year, they would delay the LHC, but they might just turn that 50% probability into a discovery. If they switched off LEP, they'd be leaving the field open to the Tevatron until the LHC started running. They chose the latter course, and at 8.00 a.m. on 2 November 2000, LEP was switched off for good. "A symposium was organised on this occasion," said Herwig, "and I was asked to give the eulogy."

Time has shown that the CERN management made the right call. Despite several years of glancing cautiously across the Atlantic, where the Tevatron experiments were inexorably narrowing the range of available masses available to the Higgs, it proved to be out of reach of both LEP and the Tevatron. The Higgs was discovered by the ATLAS and CMS experiments at the LHC, who announced the discovery on 4 July 2012. Its mass is 125 GeV. "LEP could have discovered it if more superconducting cavities had been installed," said Herwig, "but no spares were available and a new order to industry would have taken a long time."

Fig. 10.1 Left to right: Robert Aymar, Luciano Maiani, Chris Llewellyn-Smith, Carlo Rubbia and Herwig Schopper—the five CERN Directors-General who had presided over the lab through the LHC's long gestation—celebrate the machine's first beam on 10 September 2008 (©CERN, All rights reserved).

LEP's Contribution to Physics

CERN has a long history of contributions to the study of the fundamental interactions of nature, in particular, what we now know as the electroweak theory, which brings electromagnetic and weak interactions together in a single theoretical framework. Tito Fazzini, Giuseppe Fidecaro, Alec Merrison, Helmut Paul and Alvin Tollestrup set the scene in July 1958 when they published a paper showing evidence for the predicted decay of pions directly to electrons. This experiment provided an important measurement of an electromagnetic interaction, pre-dating the emergence of modern electroweak theory. It also it set a clear direction for CERN, and it made headlines around the world.

Electroweak theory was developed in the 1960s. By the 1970s it had reached theoretical maturity. With the discovery of weak neutral currents at CERN in 1973, and the W and Z bosons, carriers of the weak force, in the 1980s, CERN experiments had taken the first steps in laying the experimental foundations that underpin the theory. LEP's legacy would be to complete those foundations. It was built for that purpose, but when the machine started up in 1989, hopes were high that it might do more.

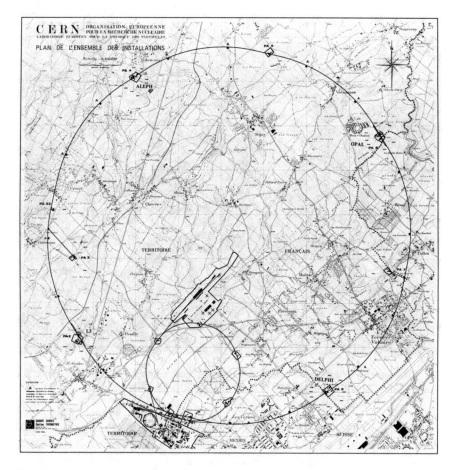

Fig. 10.2 The final position of LEP, with the sites of the four experiments marked on the map (©CERN, All rights reserved).

Theory and Experiment

"Whenever a new facility for particle physics starts up, it's a step into the unknown," said Herwig. "These machines are designed to venture into unexplored territory, which is why working with them is so exciting." LEP was no different. Designed specifically to put the electroweak theory to the test, it also ventured to higher energies than had been achieved before, and so opened up the enticing prospect of discovering something new and unexpected. "A good theory not only describes known experimental truths, it also makes predictions that can be tested," explained Herwig. "This was the case for the electroweak theory at the end of the 1980s. The job of experiments is to test those predictions to establish which theories best describe reality."

Fig. 10.3 An aerial view of CERN and its surroundings, showing the position of the LEP ring. The runway of Geneva airport gives an idea of the scale (©CERN, All rights reserved).

Even before LEP started up, however, it was already clear that the Standard Model of particle physics, of which electroweak theory is a key component, had limitations. New physics, and new theories, would be needed to explain some phenomena that had already been observed. "Sometimes something completely unexpected is found by experiments," explained Herwig, "and we were all hoping for that with LEP. Such a discovery would not have destroyed the established knowledge—there are no real revolutions in physics, just steps beyond the domain of applicability of the existing theory."

Einstein's special theory of relativity is a key example. It superseded Newtonian gravity, but it did not invalidate it. Rather, it showed that Newton's theory could only be applied at velocities that are low compared to the speed of light. Newtonian gravity is a special case of Einstein's theories. In a similar way, physicists were hoping that LEP would be the machine to show that the Standard Model is a special case of some broader theory.

"When LEP started up, we had expectations in both directions," said Herwig. "With beam energies of 50 GeV we expected to verify the Standard Model to high precision, but there was the hope, particularly with LEP II, that we'd find something new. After 12 years of hard work, we can safely say that LEP surpassed all expectations as far as verifying the Standard Model is concerned, turning it into a high-precision field. On the other hand, no spectacular new discovery was made at LEP, leaving physics beyond the Standard Model to the next generation."

In His Own Words: Incredible Precision and a Lasting Legacy

"We always refer to the big accelerators at CERN as machines, but really they are incredibly precise scientific instruments. In order to achieve the precision that it did, LEP I's beam energies had to be determined and kept constant with a precision of about 1 in 5000. This required great skill from both the accelerator teams and the experiments, and very close collaboration between the two. At the time, most experimental physicists carrying out experiments at CERN were used to turning up at the lab and simply having beams on tap. With LEP, they had to learn a whole new language. They acquired the accelerator physicists' and engineers' jargon, just as the accelerator teams learned theirs. As time went on, the experiments developed ways of working together as well, which bode well for the future. In research, it's important to keep human bias as far from the analysis as possible, since humans are very good at seeing what they want to see. That's partly why independent verification of results is a key part of any analysis. In the early days of LEP, the experiments kept very much to themselves to avoid any cross-contamination, but as time went on, they put procedures in place to combine their results when the time was right in order to get the best possible precision on the final measurement. This is standard practice these days, but at the end of the 1980s, that kind of collaboration between experiments was new.

Calibrating a scientific instrument is always a challenge, but when it's 27-km around, it becomes even more difficult. The beam energy of LEP was determined by the magnetic field along the beam path, and by the diameter of the ring. That's a fairly simple calculation, but with an instrument so big, even tiny changes in the diameter of the ring could have a significant influence on the calculated value for the beam energy. The magnetic field could be precisely measured, but the diameter of the ring was subject to factors beyond the control of even the best accelerator engineers. It was influenced by the sun, the moon, the amount of rainfall in the Geneva basin, and even the TGV trains accelerating out of Geneva's railway station on their way to Paris.

Earth tides caused by the influence of the sun and the moon, and rainfall changing the water levels within the Jura mountains and Lake Geneva could lead to variations in the diameter of as much as a metre. Once these effects were understood and implemented into the beam energy calculation, there remained one more fluctuation that took longer to understand. Its timing was regular, but not linked to any natural phenomenon that we could understand. We only solved it when someone noticed that it looked very much like the timetable for departures of the TGV to Paris, and that proved to be the cause. The TGV uses direct current electricity, which literally

Fig. 10.4 LEP's innovative so-called concrete magnets stored in the ISR tunnel awaiting installation. This was the first time the Laboratory had gone into mass production at this scale (©CERN, All rights reserved).

used the Earth as a return path. Since the tracks passed close to the LEP ring [3], the current preferred the low-resistance path offered by the LEP vacuum chamber and influenced the magnetic field along the beam path. When all these factors were understood, the residual systematic error in the beam energy for the mass of the Z particle was just 0.0017 GeV of the 90 GeV.

LEP, and above all the high precision achieved by the experiments, had established the basis for the further exploration of the standard model. For example, the mass of the top quark for which no theoretical prediction existed, could be deduced from the LEP results with high precision. The top quark was later detected at Fermilab, outside the range of masses available to LEP, but exactly where the LEP measurements had predicted it would be. Even more important for the future of CERN was the mere existence of the LEP tunnel. Although the original idea to install a hadron collider

Fig. 10.5 The spherical storage structure on top of LEP's copper accelerating cavities saved energy by ensuring that the accelerating field was applied only when beams were passing through the accelerating structures and stored at other times (©CERN, All rights reserved).

along with LEP had to be abandoned because of space limitations, the tunnel had nevertheless been designed with a hadron collider in mind. For the LHC, which replaced LEP in the tunnel, the 27-km circumference was essential. Had we only wanted to do electroweak physics with LEP, a circumference of about 20 km would have been sufficient. Despite going against the advice of many colleagues when we were designing the tunnel, insisting on the larger circumference proved the right thing to do. Without the existing tunnel, I doubt that the LHC would have been approved at all.

The great triumph of the LHC experiments so far has been the discovery of the Higgs boson with a mass within the range predicted by the LEP experiments. The LHC and its experiments, supported by amazing advances in scientific computing, have also proved their ability to do precision experiments. These have, so far,

Fig. 10.6 Jack Steinberger stands in front of the ALEPH detector. Steinberger was the experiment's spokesperson (©CERN, All rights reserved).

confirmed all expectations of the standard model, further cementing the monumental intellectual achievement that it represents. No definite signs of new physics beyond the standard model have been found, although there are some promising indications. Particle physics research is a painstaking process, a very important part of which is narrowing down the range of theories proposed to take the field beyond the standard model. The LHC is playing a valuable role in this respect. In a few years' time, it will begin a new phase, called high luminosity LHC, or HL-LHC for short. This will deliver much more data than the current LHC, improving the experiments' sensitivity to new physics. The whole community is looking forward to that.

LEP's legacy projects into the long-term future of CERN. The particle physics community has identified a future circular collider (FCC), with a circumference of

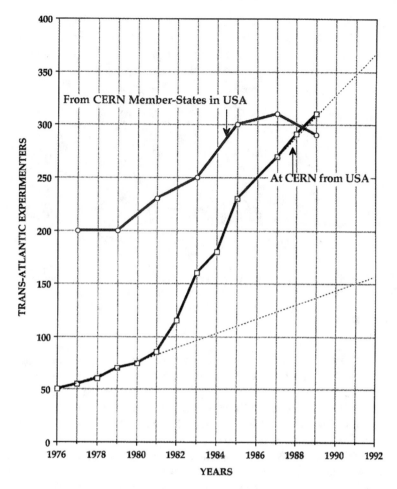

Fig. 10.7 A comparison of the number of particle physicists from CERN Member States working in the USA compared to the number of Americans working at CERN shows that there were more Americans at CERN for the first time in 1989—the year LEP started up (H. Schopper, LEP, 2009, 978-3-540-89300-4, ©Springer, All rights reserved).

about 90 km as the best facility from a pure physics perspective to take the field to the end of the twenty-first century. Such a machine would take over from the LHC in the 2040s. A study is underway with the goal of establishing whether it would be feasible from technological, geological, financial and environmental points of view. One thing that's already clear, however, is that if the FCC ever sees the light of day, it will follow the LEP–LHC strategy of first installing an electron collider, to be replaced later by a hadron machine. I think it's fair to conclude that LEP changed the course of high energy particle physics profoundly, and its legacy is still being played out.

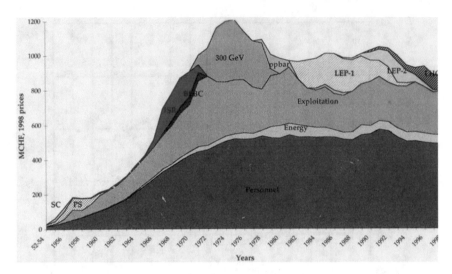

Fig. 10.8 CERN's budget rose rapidly in the early years of the Laboratory, reaching a peak in the 1970s. A constant budget was a condition for LEP's approval. The CERN budget has remained constant in real terms since then (H. Schopper, LEP, 2009, 978-3-540-89300-4, ©Springer, All rights reserved).

Over the 70 years that CERN has existed, it has achieved tremendous recognition. This is first and foremost due to its scientific achievements. But at the laboratory's foundation lies the concept of science for peace. This has allowed the positive values of science to be translated into international collaborations between countries—both for accelerators like LEP, the LHC and possibly the FCC in the future, as well as the experiments that study their collisions.

CERN and its experimental collaborations have demonstrated the benefit of diversity and inclusion: of non-discrimination between nationalities, races, religions, gender, political conviction or tradition. Over the years, CERN has grown, and maybe one day will evolve from being a European laboratory to a World laboratory by statute. Such a development would, of course, require certain changes in the Organization's convention and discussion in this direction have started already.

These days, it seems to me that CERN has become a byword for successful international collaboration. We hear people calling for 'a CERN for climate science', or 'a CERN for AI', or indeed for any number of major challenges facing humanity. Could the CERN model become a kind of template for collaboration between countries or even any new social structures that might emerge in the coming decades in our rapidly changing world? In 2023, as part of the UN-endorsed International Year of Basic Sciences for Sustainable Development, a panel discussion was organized at UN headquarters in New York. Moderated by my former colleague at CERN, Maurizio Bona, and with closing remarks from Michel Spiro, President of the International Union of Pure and Applied Physics, and a former President of the CERN Council, the session concluded with a call for international coordination of sustainability science.

With one of my successors as Director-General of CERN, Rolf Heuer, on the panel, the influence of CERN in this development is clear."

References

1. Llewellyn-Smith C (2007) How the LHC came to be. Nature 448:281–284. https://doi.org/10.1038/nature06076
2. https://cerncourier.com/a/the-w-and-z-at-lep/
3. https://cds.cern.ch/record/309231/files/sl-96-036.pdf

Chapter 11
Science for Peace with SESAME and SEEIIST

They say that the main CERN restaurant, once known as the Co-op after the company that used to run it, now more prosaically known as R1, is the place where everything of import that happens at the laboratory is discussed. It's a place where students fresh out of high school can rub shoulders with Nobel Prize winners, and over the years, it has been the scene of many important discussions.

One such conversation took place in 1995 between theoretical physicists and old friends, Eliezer Rabinovici and Sergio Fubini. A professor at the Hebrew University in Jerusalem, Rabinovici had grown up in the shadow of the holocaust and had spent much of his life seeking to promote peace between Israel and its neighbours. He was a vociferous supporter of developing education in the west bank and had co-authored a report arguing for Israeli support for the universities there. The two discussed the CERN model for peaceful scientific cooperation between nations. If it could work for Europe in the aftermath of the Second World War, could it work today in the strife-torn Middle East? One result of their meeting was the establishment of the Middle East Scientific Cooperation Committee (MESC), which was formalised later that year with a signature in Cairo.

This was not the first time that such an idea had been aired. The Pakistani Nobel laureate, Abdus Salam, is often credited as being the first to propose a regional facility. He was a strong advocate for science in the developing world, being at the origin of both the World Academy of Sciences (TWAS [1]) and the International Centre for Theoretical Physics (ICTP), which now bears his name. Both have a mission of promoting science in the developing world. In an article compiled from interviews given by Salam, and published posthumously by the ICTP in 2004 [2], he summed up his views: "Funds allotted for science in developing countries are small, and the scientific communities sub-critical. Developing countries must realize that the scientific men and women are a precious asset. They must be given opportunities,

© The Author(s) 2024
H. Schopper and J. Gillies, *Herwig Schopper*, Springer Biographies,
https://doi.org/10.1007/978-3-031-51042-7_11

responsibilities for the scientific and technological developments in their countries. Quite often, the small numbers that exist are underutilized. The goal must be to increase their numbers because a world divided between the haves and have nots of science and technology cannot endure in equilibrium. It is our duty to redress this inequity."

Herwig recalls one particular meeting in the 1980s. It was organised by the Turkish parliament, and participation was strictly by invitation only. "Abdus Salam advocated a scientific facility for the region with the argument that the Quran encourages the exploration of nature," recalled Herwig. "At the same time, I reminded the parliamentarians that the Islamic countries had a proud tradition in science: about a thousand years ago, they were world leaders. Later I visited Saudi Arabia and suggested they could establish such a facility in Riyadh, but the time was not yet ripe."

When the second Oslo peace accord was signed in September 1995, however, there was a palpable sense of optimism in the Middle East, and it led Rabinovici to think that a window of opportunity was opening. MESC organised a meeting in November at Dahab, a fishing village turned resort on the Egyptian Red Sea coast. Held in a Bedouin tent, the meeting's agenda was Arab–Israeli scientific cooperation. Egypt's science minister, Venice Gouda, was there, as was the president of the Israeli Academy of Sciences and Humanities, Jacob Ziv. Rabinovici and Fubini both addressed the assembled scientists and diplomats, advocating mechanisms to promote cross-border research in the region.

It was a promising start, but the Dahab meeting nevertheless took place under a cloud. Rabinovici, with his dry sense of humour, later recalled that Mount Sinai shook under the influence of a small earthquake as the meeting was taking place—the first of many omens that would dog the development of what would eventually become the SESAME laboratory in Jordan. That was a trivial enough occurrence, but shortly before the Dahab meeting, there had been a far more serious development in Israel. On 4 November 1995, while speaking at a peace rally, the Israeli Prime Minister, Yitzhak Rabin, was assassinated by an ultra right wing student opposed to rapprochement with Israel's Palestinian neighbours. The flame of peace had burned but shortly, and it was left to MESC to carry the embers forward.

"At a MESC round table discussion in Turin, Gustav-Adolf Voss from DESY proposed building a synchrotron radiation facility in the Middle East," explained Herwig. "He, along with Herman Winick from SLAC, were members of an advisory committee at the BESSY I synchrotron facility in Berlin, and when they learned that the machine was to be decommissioned, they suggested donating it to the region."

It was at another MESC meeting in Turin in November 1997 that this idea took wing. "It was a seminar attended by about 30 scientists from Israel and Arab countries," recalled Herwig, "and they decided to set up a steering group to take the idea forward." Sergio Fubini asked Herwig, with his long international experience, to chair that group.

The New Laboratory Takes Seed

Herwig accepted the challenge on condition that the CERN model be adopted, which implied that UNESCO would play a key role in steering the proposed laboratory into existence. Among the first things he did was confirm that BESSY I was indeed scheduled to be decommissioned. "Persuading the German authorities to donate BESSY I, as suggested by Voss and Winick, was relatively easy," recalled Herwig, "they agreed to this at the end of 1999, but they made two conditions: first, it had to be done immediately since the building had been promised to another organisation, and second, it had to be done without cost for Germany." This was a start, but at the end of the 1990s, there was no receiving organisation and no budget. Solving these issues became Herwig's immediate goal.

In January 1998, Herwig persuaded Sergio Fubini to sign a letter with him to UNESCO's Director-General, Federico Mayor, requesting that UNESCO provide an umbrella for the proposed laboratory as it got off the ground. They received an enthusiastic response.

"As a next step, it was important to find out whether there was sufficient interest among the scientists of the region, or whether such a facility would become a white elephant," recalled Herwig. Another MESC workshop was organised in April 1998 to find out. Tord Ekelöf, a Swedish CERN physicist was the organiser, and he held the workshop at his home institute of Uppsala University, with a visit to the Swedish Academy of Sciences. The conclusion was that there would be no shortage of users for such a facility in principle, but that details of how it would be implemented should be worked out as soon as possible.

Things had got off to a promising start, but the premonitions that Eliezer Rabinovici felt at that meeting in Sinai proved to be well founded: there would be many obstacles to be overcome before SESAME came into operation. The laboratory had a name, the hardware to build a machine, a fledgling user community, and it had the support of UNESCO. Still missing were members, if it was to follow the CERN model, a physical location, and beamlines to host experiments. The fact that BESSY I was a second generation light source at a time when third generation machines were becoming established would also prove to be an issue: by the time BESSY I was installed in its new home, would it still be attractive to users? Herwig was taking things one step at a time.

UNESCO Takes the Lead

Federico Mayor's first action was to call a meeting at UNESCO's headquarters in Paris to which governments from all around the Mediterranean and the Middle East and North Africa (MENA) region were invited. The meeting took place in June 1999. It established an interim council (IC), for SESAME, in much the same way that a far earlier UNESCO meeting had established the Conseil Européen pour la Recherche

Fig. 11.1 Swedish physicist Tord Ekelöf raises a toast during a MESC workshop he hosted at Uppsala University in April 1998. The workshop concluded that there would be no shortage of demand for a lightsource in the Middle East (Herwig Schopper's personal collection. ©Herwig Schopper, All rights reserved)

Nucléaire (CERN) in 1952. At its first meeting, the IC appointed Herwig to be its chair. The IC's job was to prepare a proposal for the establishment of an international organisation in the Mediterranean region to be submitted to a plenary meeting of UNESCO. "This event should be considered as the conception of SESAME, if not its birth," explained Herwig. "And Federico Mayor was certainly one of the founding fathers."

"After the pioneering role that Sergio Fubini had played, he was taken ill and was unable to pursue the dream he'd helped to initiate any further," said Herwig. "That meant that preparing the proposal for a new international laboratory fell largely to me." In 2005, Fubini passed away before seeing SESAME reach fruition. Today, researchers visiting the laboratory to carry out their experiments are reminded of his role through the Sergio Fubini guest house, which provides accommodation and meeting facilities for visiting scientists, and was funded by the Italian government.

At the June 1999 UNESCO meeting, Algeria, Armenia, Bahrain, Cyprus, Egypt, Greece, Iran, Israel, Jordan, Morocco, Oman, the Palestinian Authority and Turkey were represented. In addition, Algeria, Sudan, the United Arab Emirates and Yemen also expressed interest in the project, and several countries from beyond the region said they would help. A series of IC meetings ensued over the following years, moving from Paris to places like Yerevan, Cairo and Amman.

In parallel, meetings were ongoing to secure the donation of BESSY I, and it was at one of these, held in Berlin in August 1999, that SESAME got its name. "I insisted that the name should have some meaning and not be a meaningless, difficult to remember, acronym," remembered Herwig. "After some discussion, the name SESAME was proposed and accepted since the word is widely known from the tales from the thousand and one nights across Arab and western cultures to mean door opener." Only later were the words Synchrotron-light for Experimental Science and Applications in the Middle East tailored to match. It's good that an appropriate acronym was found, but the door opener became an important psychological element in defending SESAME in political and public spheres."

Securing BESSY

When the IC met in Paris in December 1999, the members of the new organisation faced their first major hurdle. SESAME was not the only bidder for BESSY I: Spain and Poland had also expressed an interest in the second generation machine. The German government favoured the Middle Eastern initiative, but Herwig still had to find a way of covering the cost of careful dismantling and documentation, as well as shipping the machine to its final destination, wherever that might be. For such a complex machine, this was not a trivial task, and it required experience. "There was a proposal to engage experts from Novosibirsk and Armenia, who knew the facility well," explained Herwig, "but they would have to be paid, and the cost was estimated at $600,000." SESAME had to find a way to guarantee this before the end of 1999 because the BESSY building had already been promised to the Max Planck Society.

"At a meeting of the IC in Paris I asked for voluntary contributions, but delegates to such meetings rarely have the power to commit funds during the meeting," said Herwig. Nevertheless, he managed to secure $200,000. During the meeting he turned to Koïchiro Matsuura, who had taken over from Federico Mayor as Director-General of UNESCO on 1 January 2000. "Matsuura invited me to lunch," he recalled, "and as we sat down to our first course, I warned him that this might become a quite expensive lunch. I asked him for the missing $400,000, without which SESAME would be dead." Much to Herwig's great surprise, and relief, Matsuura's enthusiasm for the project matched that of his predecessor, and he agreed on the spot to provide the necessary funds. This ruffled some feathers: certain ambassadors to UNESCO later complained that Matsuura had not followed established UNESCO protocols, but in reality, the Japanese government had provided him with a considerable sum to be used at his discretion when he became Director-General.

SESAME's immediate future appeared to be secure as the Russian and Armenian experts got to work, but there would soon be further hurdles to cross. Before long, there was a request to the German government to withdraw the export authorisation for BESSY I because, according to some, it could be used to produce fissile material for atomic weapons. "I was invited to a discussion on German TV with a chemist from Marburg called Brandt," said Herwig. "I had to admit that SESAME could produce small numbers of atoms of uranium, or even plutonium, but not enough to make a weapon. I pointed out that if synchrotrons could do that, they would also be able to produce quantities of gold, and in that case, SESAME would not have any financial problems at all." Despite the high-profile nature of the case, it was quickly resolved in a rational way, and BESSY I left the port of Hamburg on a container ship bound for Jordan on 7 June 2002.

The Formal Foundation of SESAME

For a new international organisation to be established under the auspices of UNESCO, the approval of the 195 governments represented in the UNESCO General Conference is required. It's a tall order. Even to get to the General Conference, which takes place every two years, a proposal has to pass through the Executive Committee, with more than 50 UNESCO members represented. Although convinced that going to UNESCO had been the right thing to do in order to guarantee the sustainability of SESAME, Herwig recognised that the approval process could be a long one. He started by revisiting the CERN convention: the international agreement under which CERN was established in 1954. "Since the CERN convention has proved itself as an extremely successful basis for international cooperation combining the objectives of conducting excellent science and bringing people together," he explained, "I more or less copied it for SESAME, with just a few minor changes." In this way, the SESAME convention stipulates one vote per member, regardless of size. Each member can send two delegates to the Council, one representing government, the other the scientific community. One amendment was that UNESCO's Director-General would be represented, with full voting rights, giving UNESCO a stronger role than at CERN, and enabling it to mitigate in case of political disagreement. "Right from the start," recalled Herwig, "the same spirit I had become used to in the CERN Council was established: delegations always strove for unanimity in decisions and looked for alternatives if any member could not agree."

When Koïchiro Matsuura became Director-General of UNESCO, he appointed the assistant-Director-General, Walter Erdelen, as his representative at the IC, and the Polish biologist, Maciej Nałęcz, as its secretary. Nałęcz was soon promoted to the position of Director of Basic Sciences and Engineering at UNESCO, and soon took UNESCO's seat on the Interim Council. "Clarissa Formosa Gauci, Assistant Programme Specialist in UNESCO's Division of Basic and Engineering Sciences, assumed the role of secretary," said Herwig, "and she has fulfilled the role untiringly ever since."

One day, shortly after Herwig had drafted a convention for SESAME, Nałęcz took him to see UNESCO's legal advisor. "Without any pleasantries, he broke into a strong criticism of the draft," recalled Herwig, "and when Maciej asked him if he knew of any better examples, he showed us the CERN convention." A few awkward moments

Fig. 11.2 Left to right: Herwig Schopper, Chris Llewellyn Smith and Maciej Nałęcz, who was Director for Basic and Engineering Sciences at UNESCO in the early days of SESAME, visiting SESAME magnets under test at CERN in 2015 (©CERN, All rights reserved)

Fig. 11.3 Two key women in the early days of SESAME are Council Secretary, Clarissa Formosa-Gauci (left), and Zehra Sayers, who Chaired the Scientific Advisory Committee (Nuovo Cim. 40, 199–239 (2017) [3] ©Springer, All rights reserved)

ensued, during which Nałęcz introduced Herwig as a former CERN Director-General and pointed out that the draft was heavily based on the CERN convention. It was rapidly approved.

There then ensued discussion on how to refer to SESAME's members. Unlike CERN, which refers to member states, not all of SESAME's potential members are recognised as states by all the others. "I was sitting in an office in Amman connected by one phone to Yasser Arafat and by another to the ministry in Jerusalem," recalled Herwig, "after long deliberations, both sides agreed to refer to the Palestinian Authority on behalf of the PLO, so that's want went into SESAME's statutes." The problem was that although Palestine was in common use at the United Nations, it had not been adopted by UNESCO due to Israeli opposition. Years later, when the SESAME council was presided over by Rolf Heuer, the third former CERN Director-General to hold the position, the term Palestine would be quietly adopted.

With the statutes approved by all of SESAME's prospective members, it was time for Herwig to seek UNESCO approval. "After considerable lobbying to accelerate the procedure, the Executive Committee recommended that the November 2001 General Assembly approve SESAME in principle, and delegate the Executive Committee to take the final decision without waiting for the next General Assembly to come around," explained Herwig. As a result, SESAME's statutes were approved in May 2002, with the Executive Committee clearly stating its aspirations for the new international organisation. They issued a statement referring to SESAME as a "quintessential UNESCO project combining capacity building with vital peace building through science [4]," and suggesting that it may serve as a blueprint for similar projects in other regions. "It was a miracle that this approval was obtained in under two years," said Herwig.

As with CERN before it, SESAME would come into legal existence once a certain number of member governments, six in the case of SESAME, had confirmed their intention to join in writing to the Director-General of UNESCO. This milestone had been reached by 6 January 2003, the day that the SESAME Interim Council became the SESAME Council. Bahrain, Egypt, Israel, Jordan, Pakistan, the Palestinian Authority and Turkey were the founding members, soon to be joined by Cyprus and Iran. "Along the way, many more countries had expressed interest in SESAME," said Herwig, "and I hope that with time they will come on board. SESAME has also benefitted from the formal support of many observers." At the time of writing, SESAME's members are Cyprus, Egypt, Iran, Israel, Jordan, Pakistan, Palestine and Turkey. Brazil, Canada, the People's Republic of China, CERN, the European Union, France, Germany, Greece, Italy, Japan, Kuwait, Portugal, the Russian Federation, Spain, Sweden, Switzerland, the United Arab Emirates, the United Kingdom, and the United States of America are observers.

In parallel with pushing the statutes through UNESCO, the SESAME Interim Council had also been hard at work to find a site for the new laboratory, and after careful deliberations, and a considerable dose of serendipity, Jordan had been chosen at a restricted meeting of the Interim Council at CERN on 10–11 January 2000 (see this chapter, In his own words: Finding a home for SESAME).

Fig. 11.4 Herwig sits front and centre for a group photo taken at the first meeting in Amman of the SESAME Interim Council in August 2001 (Herwig Schopper's personal collection. ©Herwig Schopper, All rights reserved)

Building the Laboratory

Once the site had been chosen, no time was lost in setting up the laboratory. A directorate had to be appointed and staff hired. Jordan's Minister of Education, Professor Khaled Toukan, became Director. A physicist by training and former president of Al-Balqa University, Toukan remains SESAME's Director to this day. Over the years, he has held various positions in the Jordanian administration. "His close links to the Royal Hashemite Court and the government of Jordan have proved invaluable on many occasions," said Herwig, "they remain essential for the good development of the laboratory." The Director is supported by an Administrative Director, the first of which was Hany Helal, an Egyptian who strongly shaped the SESAME administration and later he became a minister in the Egyptian government. This post has so-far been filled exclusively by Egyptians. A Technical Director, hired from the European synchrotron light source community, to bring the necessary expertise to the project completed the initial top management team. Later, when the facility was approaching completion, a Research Director was appointed, again bringing the necessary expertise from Europe.

A training programme was also set up with support from the International Atomic Energy Agency, whose origins can be traced to Eisenhower's landmark 'Atoms for Peace' address to the United Nations General Assembly in 1953. Its goal was to

Fig. 11.5 Four key SESAME figures during a meeting at UNESCO in Paris. Left to right: Khaled Toukan, Director of SESAME, Eliezer Rabinovici, SESAME Council delegate, Zehra Sayers, Chair of the SESAME Scientific Advisory Committee, and President of the SESAME Council, Herwig Schopper (Herwig Schopper's personal collection. ©Herwig Schopper, All rights reserved)

ensure that SESAME would be able to recruit staff competent in all the skills needed to operate a world-class laboratory from its members, and in this it has been a resounding success: since the start of operation, SESAME has been operated largely by staff from its members.

Another important task was to provide a building for the laboratory. The Jordanian offer to host SESAME came with a promise to provide a site in Allan, north of the capital Amman, along with the necessary resources for construction (see this chapter In his own words: Finding a home for SESAME). The next task was to design an appropriate building.

To save time, a decision was taken to copy the building housing the Angströmquelle Karlsruhe, ANKA, light source at the Forschungszentrum Karlsruhe, FZK. "We appointed Dieter Einfeld to be SESAME's first Technical Director," explained Herwig." As someone who knew ANKA well, he had a good feel for the building already." FZK made the plans for the building available. They were translated into Arabic, and a local architect was engaged to adapt the building to the local context. Soon, everything was ready for construction to begin, and the groundbreaking ceremony was held on 6 January 2003, with H. M. King Abdullah II and Koïchiro Matsuura unveiling a commemorative plaque at the site in Allan.

Even though the building was tried and tested, it came with surprises. The first was that the quality of the ground required foundations deeper than foreseen in order to guarantee the stability required for a synchrotron. "The host state Jordan honoured its

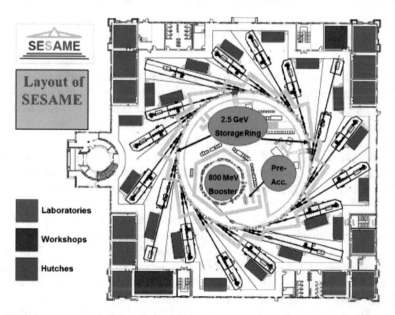

Fig. 11.6 The original layout of the SESAME laboratory showing laboratories, workshops and experimental hutches. The synchrotron is in the centre of the building (©SESAME, All rights reserved)

Fig. 11.7 The second SESAME users' meeting took place in Isfahan in October 2003 (Nuovo Cim. 40, 199–239 (2017) [3] ©Springer, All rights reserved)

commitment to cover the cost," said Herwig, "but a delay in construction could not be avoided." Nevertheless, five years later in 2008, the building was ready to welcome the new facility and was inaugurated, again in the presence of Koïchiro Matsuura and H. M. King Abdullah II. "On this occasion," recalled Herwig, "I handed on the Presidency of the Council to Chris Llewellyn-Smith, but I retained a close interest in SESAME's development." Herwig had served for ten years and was made an honorary member of the SESAME Council for life. Like Herwig, Llewellyn-Smith had been Director-General of CERN. His task as President would be to see the facility through to full operation.

Things started well as BESSY I was transferred to the new building. By this time, it had become clear that simply rebuilding BESSY I would not attract good scientists to the lab, and ambitious plans had been hatched to build a third generation light source instead, using components from BESSY I as the injector. Although funding for such a machine was still under discussion, the building was configured to accommodate a modern machine, and the BESSY I components were installed as an 800 MeV injector consisting of a machine called a microtron, and a ring made up of BESSY I components that was named the booster.

Commissioning was underway when disaster struck. Despite having been designed for a European climate, a particularly harsh winter hit Jordan in 2013, and on 14 December, the building's roof collapsed under the weight of snow and ice. "Luckily the booster was saved by its shielding," said Herwig, "so with the addition of some ad hoc plastic waterproofing, commissioning could continue." Jordan remained true to its promise, and a deal was struck whereby the building contractor would pay for two-thirds of the cost of a new roof, with the host state covering the rest. Before long, a new roof was under construction. In the meantime, the microtron and booster were fully commissioned, with the booster achieving its design energy on 3 September 2014, under an open sky. The new roof was completed the following year.

Towards the Third Generation

Even before BESSY I had been secured for SESAME, wheels had been set in motion that would lead to the donated machine being used as a catalyst for a modern third generation light source in the Middle East. The first generation of such facilities had largely been run as parasitic experiments at machines built for particle physics at laboratories such as DESY. The second generation consisted of machines dedicated to synchrotron radiation research, and the third includes so-called insertion devices that are inserted into the ring between bending magnets to produce beams with specific characteristics that allow them to extend the range of research that can be done. BESSY I was a second generation machine, and an early one at that.

"At the Berlin workshop in August 1999, an international study group was set up to work out a first proposal for what we might do with BESSY I," explained Herwig. That group delivered a proposal in October of the same year to increase the energy from 800 MeV to 1 GeV by increasing the diameter of the ring from BESSY I's 62 m to just over 100 m. Although this proposal showed the serious intent of the programme, objections were soon raised that given the emergence of third generation machines, SESAME's role would be limited to training unless a more ambitious plan were adopted: a second generation machine would simply not be competitive for research.

Under the guidance of Technical Director, Dieter Einfeld, a more radical proposal, SESAME II, was advanced in April 2002. This was the first time that a new main ring was proposed. A third generation machine that would use BESSY I as an injector, the new look SESAME would be a 120 metre circumference third generation facility with an energy of 2 GeV. "The cost estimate was about $14 million," said Herwig. "It seemed unlikely that we could obtain this from the members, so we had to develop a somewhat unusual finance policy." A call was put out to light source laboratories around the world for donations of equipment, and a request was submitted to the European Union.

Over the coming years, the design evolved until it reached SESAME IV. This was for a 2.5 GeV machine, 133.2 metres in circumference with 50 metres of straight sections that could incorporate insertion devices. "These parameters would make SESAME a fully competitive third generation facility," said Herwig, "and the only one of its kind in the region." All that remained was to find the funding.

"The chances of finding funding from the members seemed extremely small," said Herwig, "but the European Union had a programme to support projects in neighbouring regions. In 1999, informal contacts had been established between UNESCO and the European Commission to sound out potential." One particularly promising avenue went by the name of MEDA [5], a programme to support countries in the Mediterranean region to reform their economic and social structures and mitigate the social and environmental consequences of economic development. MEDA had a budget of €3.4 billion over three years. "After a meeting in Brussels where we learned about the programme, we submitted a request to the commissioner for external relations, Chris Patten, on 23 July 2001 requesting 10 million Euros for a new main ring," said Herwig. The initial signs were good, with the research ministers of France and Germany supporting the idea, along with the commissioner for research. Nevertheless, there was a long road to travel before funding would be secured.

In October 2001, an EU delegation attended the SESAME Interim Council meeting, and explained that an independent evaluation of the SESAME proposal would have to be undertaken. Before that could happen, a competitive adjudication process to establish who would carry out the evaluation was necessary. An international committee was sought. By the time the whole process was over, the summer leaves of 2002 were starting to fall. The resulting report raised as many questions as it answered and threw the ball back into Herwig's court. "In August 2003, SESAME submitted several documents to the EU answering all the questions," explained Herwig, "and when we eventually received a reply, it just said that the commission was not able to provide community funding to SESAME."

Herwig did not give up, and his persistence eventually paid off. "The people from the Commission's neighbouring regions programme were not accustomed to negotiating with scientists," he explained, "and they insisted on dealing only with the Jordanian government." As a consequence, invitations and documentation were sent to the Jordanian Embassy in Brussels, translated into Arabic and sent on to Amman, where they were translated back into English and sent to Herwig. "When I saw the names of the people that were invited to the meeting in Brussels, they were all unfamiliar to me," he explained, "so imagine my surprise when I found that I knew almost everyone in the room." It seems that the translation and retranslation of names had corrupted the spellings beyond recognition.

The meeting took place on 10 March 2004, and discussions took a surprisingly technical turn. "People argued that in order to be competitive, the machine would need higher energy than 2 GeV," recalled Herwig, "so that's when we came up with the plan that would allow the machine to go to 2.5 GeV, which brought the budget up to 12 million." All the commission's remaining questions were easily dealt with, but there were still hurdles to cross. There was discussion on which funding programme to use, and the meeting closed with an agreement that an initial grant of €1.2 million could be made under a bilateral programme with Jordan, should the Jordanian government request it. "The following extended discussions with the EU kept coming back to the same position," said Herwig. "The project is good and the financial needs acceptable, but it was not clear which EU programme should be used to support it."

"This was a useful start," said Herwig, "but it was not enough. We had put together a funding model whereby each member would pay according to their ability, but this was for operational costs and not capital investment. When I handed over the presidency to Chris Llewellyn-Smith, we still had not found a solution for funding the main ring."

A major step forward was made in 2012 when Iran, Israel, Jordan and Turkey each committed to making voluntary contributions of $5 million. CERN also came on board at this time. Thanks to the engagement of CERN's Director-General, Rolf Heuer, the EU's seemingly insurmountable hurdle was overcome by providing the funds to the European laboratory rather than to SEAME directly. On 23 May 2013, the commission and CERN announced the CESSAMag project, whereby CERN would oversee the design and production of the magnet system for the SESAME main ring. Finally, there was light at the end of the tunnel.

On 1 February 2016, the last components of the magnet system left CERN for Jordan. Beamlines had been donated by light sources around the world, and Italy made substantial contributions in the form of the accelerating system for the new machine and the Sergio Fubini guesthouse. On 12 January 2017, the main ring circulated its first beam, and on 16 May 2017, the laboratory was formally opened. "After a successful period at the helm, Chris passed the presidency of the SESAME council to a third former CERN Director-General, Rolf Heuer," said Herwig, "and the research programme got underway. Today, SESAME is fulfilling the initial vision of producing world-class research and bringing people from across the region together." Europe continued to support SESAME, providing funds to the Jordanian government that were used to build a solar power plant, making SESAME the first large research infrastructure in the world to be powered by renewable energy.

SESAME got off to a great start and is becoming established as an important player on the international scientific stage. However, Herwig is worried for its long-term future. "It is a pity that a jewel, both from the political and from the technical point of view, is suffering because of lack of funds for full operation," he explained. "Mainly because of the complicated political situation, some members are not able to pay their full yearly contributions to establish a modest budget. It is thanks to the help of some observers that a research programme can be maintained. Let us hope that the situation improves, and perhaps more countries can join this unique international project, giving it the financial stability it needs."

Fig. 11.8 The first three presidents of the SESAME Council. Left to right: Chris Llewellyn-Smith, Herwig Schopper, Rolf Heuer (Nuovo Cim. 40, 199–239 (2017) [3] ©Springer, All rights reserved)

The Origins of SEEIIST

In autumn 2016, Herwig attended a meeting in Dubrovnik organised by the World Academy of Art and Science [6]. There was an item on the agenda about development in the Balkans, but discussions were going nowhere. "After a sleepless night," he recalled, "I came to the conclusion that a project like CERN or SESAME would be the right initiative since the region needed both science and technology, and an improvement in relations between governments. The same night I drafted a proposal."

Herwig gave his proposed initiative the name South East European International Institute for Sustainable Technology (SEEIIST). "I gave it a flexible title since it was not clear which kind of project might be chosen," he explained. "It was discussed extensively and adopted with a certain enthusiasm. I hoped that the clumsy title would be changed when a concrete project was chosen, but it is still used today."

Herwig had in mind a large project that would require cooperation between several countries, providing training opportunities for young people from across the region, while, like SESAME, encouraging scientists to work across borders, to stay in the region, and to reverse brain drain from the region.

After the Dubrovnik meeting, several projects were put forward, but none gained sufficient traction to move the idea forward. The region has a turbulent past, and it proved hard to find a project that SEEIIST's proponents felt they could push at the level of national governments. "I was close to giving up," recalled Herwig, "but then by chance I met Sanja Damjanović in March 2017. She was a physicist participating in an experiment at CERN and had recently been appointed Science Minister in Montenegro." Damjanović became Herwig's staunchest ally. "She introduced me to Montenegro's Prime Minister, Duško Marković," he continued, "who appointed me as a kind of advisor." Following this breakthrough, a meeting of the research ministers was arranged. It had to take place at a neutral place in order that all participants be on the same footing. "I am very thankful that CERN Director-General Fabiola Gianotti agreed that we could hold the meeting at CERN."

On 25 October 2017 ministers or their representatives from ten countries met at CERN. As the originator of the idea, and as someone with years if not decades of similar experience, Herwig was asked to chair. Initial worries that representatives of the various countries would be uncomfortable talking to each other proved unfounded. "I think this was thanks to the unique ambiance of CERN," said Herwig. "I was able to expound my vision of creating a project on the model of CERN, with the double mission of promoting science, technology and industry while at the same time bringing countries of the region together: building trust to mitigate tensions between countries in the spirit of science for peace."

The meeting concluded with a declaration of intent signed by Albania, Bosnia and Herzegovina, Bulgaria, Kosovo (with a footnote taking into account the reservations of Serbia), the former Yugoslav Republic of Macedonia (now the Republic of North Macedonia), Montenegro, Serbia and Slovenia. Croatia had agreed also, but for formal reasons had to delay the signature. Greece participated as an observer. In the now time-honoured tradition, a steering committee was established on the basis of one member, one vote. Despite being invited to chair that committee, Herwig declined, preferring to leave the task to someone younger. "I agreed to be a member of the committee for life, without vote" he said, "and Sanja Damjanović was elected as chair. Until this day, that declaration remains the only legal basis for SEEIIST."

Two Concrete Studies

Before the meeting at CERN, Herwig established two study groups to examine the potential for accelerator-based facilities in the region. "One possibility we considered was a 4th generation synchrotron light source," he explained, "I had good contacts through SESAME, and Dieter Einfeld agreed to be coordinator." The second group investigated the potential for a state of the art cancer therapy and biomedical research facility. This group was coordinated by Ugo Amaldi, who had worked for many years at CERN and had spearheaded an initiative that led to the foundation of a similar facility, CNAO, in Pavia, of which he was the first director. The accelerator at CNAO was based on designs produced at CERN, which continues to play an active role in developing accelerator technologies for such applications.

"In order to find out which of the two options the scientific community of the region would prefer, I proposed a forum where both would be presented," said Herwig. "We again needed a neutral location, but this time closer to the region, so we went to the International Centre for Theoretical Physics (ICTP), in Trieste, with which I had good relations since its foundation under Abdus Salam." The forum took place on 25 and 26 January 2018 and attracted over 100 participants. Both projects were well received, with neither emerging as a clear favourite. After the forum, the participants were offered a visit to Trieste's Elettra light source, where they got a foretaste of what might be.

"As often happens," commented Herwig, "the crucial event took place outside the official agenda. Before pursuing the SEEIIST project I had met the European commissioner responsible for research and innovation, Carlos Moedas. He attended the inauguration of SESAME in 2017, and I asked him whether a similar project for south-east Europe would have a chance of EU funding."

Moedas answered positively, which emboldened Herwig to persist. Moedas was invited to attend the ICTP forum. "Unfortunately, the Commissioner could not come but Robert-Jan Smits was there," recalled Herwig. Smits was Director-General for research and innovation at the Commission: a sign of the importance that Brussels attached to the initiative. "I invited Smits and his aid Bernhard Fabianek to a private lunch with Sanja Damjanović and me," recalled Herwig. "We explained that the project could only continue if we could get some seed money quickly, and when I suggested the figure of a million Euros, Smits asked whether three might be more useful." SEEIIST left that lunch with an immediate promise of one million Euros for a start-up phase of about a year, with up to three million if that phase proved to be a success. "That brave decision saved the whole project," said Herwig.

Over time, a majority of SEEIIST members came down in favour of the cancer therapy and research centre, and the idea of a 4th generation light source was abandoned. "In March 2018, the steering committee voted unanimously for this option," said Herwig, "stipulating that half the beam time should be for treatment and half for bio-radiological research." Such a research programme would be an important addition to the European research landscape, with relevance not only for developing cancer therapy, but also in areas as diverse as understanding the impacts of cosmic and terrestrial radiation, and preparing for future human spaceflight, in which radiation is an important factor. "It turned out that the visit to Elettra had made an impact on the forum's delegates," said Herwig. "They were surprised by the complexity of the facility and questioned whether the region had a sufficiently large potential user community." On the other hand, each of the potential SEEIIST members had hospitals with some expertise in cancer treatment, so the potential user community was already quite mature. "These arguments were certainly valid and had to be accepted," said Herwig, "but personally I regretted the decision since in the long run a synchrotron light source would have contributed to a much broader field of research."

Dieter Einfeld wound up the synchrotron study group, which published a record of its work as a CERN yellow report [7] in 2019. Efforts then focused on designing a radiation facility that would be unique and complementary to existing facilities. Ugo Amaldi's study group published its work as a CERN yellow report [8] in 2020, and this formed the basis of a preliminary proposal for inclusion on the European roadmap for research infrastructures.

The European Strategy Forum on Research Infrastructures (ESFRI), the custodian of the roadmap, welcomed the concept, but asked for evidence of sustainability in the form of firm commitments from potential future member states.

Phase 1 of SEEIIST

In April 2018, just after the SEEIIST steering committee had thrown its weight behind the cancer therapy and research centre, Robert-Jan Smits' term as Director-General for research and innovation came to an end and he was succeeded by Jean-Eric Paquet. "When we met him to discuss the legal implementation of phase 1, Paquet had only been in office for a few days," explained Herwig, "and he was unfamiliar with the project. This led to some confusion, but we convinced him, and he was extremely helpful."

As SEEIIST had no legal status at the time, establishing a mechanism for deploying the funds that Smits had promised was a complex affair. For such cases the European Commission had already chosen an agency attached to the German Aerospace Center (DLR) in Jülich to be an intermediary with the responsibility for ensuring that the money be spent according to existing legal frameworks. DLR asked the steering committee to propose existing laboratories to manage the expenditures. "CERN was willing to take responsibility for the accelerator part of the project but not for the rest," said Herwig. "So we had to act fast to find partners to look after the research angle, and to manage the expenses that would be made in the SEEIIST region." In the end, GSI in Darmstadt agreed to take care of the scientific part while the Slovenian Academy of Engineering would handle local expenses for the region. "The choice came down to who we knew," said Herwig. "Sanja had worked at GSI during her PhD studies and knew people there, and the president of the Slovenian Academy, Mark Plesko, was well known at CERN since he ran an accelerator controls company."

The funds were distributed evenly between CERN, GSI and the Academy. "This turned out to make daily life very difficult," explained Herwig, "since any shift of funds required an amendment to the original contract to ensure there was no wrongdoing. After some discussions, an association was established in Geneva under Swiss law to manage the funds for phase 1 but it came rather late and phase 1 had to be handled mainly by the three partners." At around this time, elections in various countries led to changes in the steering committee, whose co-chairs became Leandar Litov from Bulgaria and Mimoza Ristova from North Macedonia.

The Swiss Initiative

The declaration of intent by the research ministers of the region looked promising, but it did not carry sufficient legal or political strength to get the project off the ground. "There was no tradition of shared international projects in science," explained Herwig, "and it soon became clear that I would have to find an umbrella organisation to play the role that UNESCO had during the gestation of both CERN and SESAME." Herwig's first thought was the EU, but when that turned out not to be possible, he turned to Switzerland. "I remembered the role the Swiss confederation has played for over a century in helping to settle political issues," he explained. "I spoke to the Swiss Foreign Minister, Ignazio Cassis, in summer 2019, and during a lunch in Lugano I asked him whether he would be prepared to help SEEIIST in this precarious situation as far as science diplomacy was concerned." Herwig knew that Cassis had an interest in science policy, and after the minister had discussed the question in Berne with other members of the government, was pleased to receive a positive answer. "Cassis asked me what the next urgent steps would be," continued Herwig, "and I explained that help was needed in two areas: establishing a convention for the new laboratory and developing a procedure for the selection of a site."

Herwig's overture bore fruit. Cassis tasked several of his team to attend SEEIIST meetings, and having convinced himself that the initiative was worthy of support, he wanted to be assured of solid interest from potential members. He proposed to invite the foreign ministers of states represented in the steering committee to a meeting in Bern. Invitations were sent for a date in September 2020, but due to COVID-19, it was postponed until 12 months later.

Herwig had kept the SEEIIST steering committee fully informed throughout of his approach. "The steering committee had given me the green light to approach Switzerland," he explained, "and Switzerland's agreement to help met with unanimous approval, and a certain degree of enthusiasm."

"The Bern event became a milestone in the history of SEEIIST," said Herwig. "Ten countries were represented either by the foreign minister, another minister or a representative. Cassis asked them whether their governments were interested in the project and the answer from all 10 countries was a resounding yes. This was the first time that official government representatives had given such a clear statement." Also attending the meeting was the European Commission's deputy commissioner for science and innovation, Signe Ratso. "This gave a strong signal that EU might be prepared to provide the funding for the initial investment," said Herwig. "She said that the newly-established EU neighbourhood initiative for the Western Balkans had been designed to support infrastructure in the region, and that in principle SEEIIST could benefit. However, she also pointed out that at least some individual countries must commit themselves to the sustainability of the new laboratory."

It was time to establish new working groups: one to work out a proposal for a convention of SEEIIST, and another to propose a procedure for the selection of the site. "I pointed out that the members of these groups should not only be competent

Fig. 11.9 Swiss Foreign Minister, Ignazio Cassis (front, 4th from left) convened a meeting of Ministers from South East Europe in Bern on 13 September 2021 to discuss SEEIIST. Invited observers included CERN Director-General, Fabiola Gianotti and Signe Ratso, Deputy Director-General of Research and Innovation at the European Commission (Credit: KEYSTONE/Alessandro della Valle, All rights reserved)

to do the job," said Herwig, "but should also be able to express the political will of their governments."

Following the Bern meeting, things started to move. The working group on the convention met the very next day, and the group looking at site selection agreed to begin work in 2022. "I was not always sure about the success of SEEIIST," said Herwig, "but following the Bern meeting, I think that the chances of success were rather good." At the time of writing, the concept is recognised as a potential flagship for the region, funds for the investment could be made available from the EU, and all the necessary formal steps are being prepared.

The Swiss foreign ministry organised the two working groups and Ambassador Stefan Estermann was appointed to chair both, later being replaced by Ambassador Alexandra Baumann. "A whole team at the ministry was efficiently helping to advance the SEEIIST project," recalled Herwig, "and the very useful coordinating role of Niccolo Iorno is much appreciated." In spring 2023 the working group for the convention, after extensive discussions with the EU, presented a draft combining the positive elements of the CERN convention with the legal formalities of a European research infrastructure consortium [9, 11]. "Some details remain to be clarified but the hope is that the statutes of SEEIIST might be approved before the end of 2023 by at least some of the countries, leaving it open for others to join later," said Herwig. "I have suggested that a high-level meeting of government representatives could formally establish the SEEIIST project by committing themselves to the statutes, thereby establishing SEEIIST legally as an international organisation. Such a status would immediately make it possible for SEEIIST to apply for EU horizon funding covering the transition period necessary to setup a council, select a main site along with other sites for planned hubs, and appoint a key staff with a director who would submit a final proposal."

"Whether SEEIIST will be able to play a similar beneficial role for South-East Europe as SESAME played for the MENA region remains to be seen," said Herwig. "To convince the governments that a common project offers additional value compared to a national project is not trivial given the recent history of the region, but promoting peaceful collaboration was, of course, one of the original motivations for SEEIIST. The project is now offered to the region on a silver platter, and it is up to the individual countries to decide whether they want to come to the table. I hope they accept. It seems to me that international collaboration is more necessary now than ever."

In His Own Words: Finding a Home for SESAME

"At the same time as developing SESAME's statutes with UNESCO, we were also looking for a site for the laboratory. As a first step the Interim Council agreed on a set of criteria that had to be satisfied by the host. Among these, the most important, but also the most difficult to fulfil, were the conditions that the laboratory must be accessible to all scientists from all over the world, and that the host should make a special contribution consisting of the land and the resources to build the laboratory's main building on it.

We received a total of 12 site proposals from seven members: Armenia, Egypt, Iran, Jordan, Oman, the Palestinian Authority and Turkey. In September 1999, I visited Egypt, Jordan and Palestine with Maurizio Iaccarino, UNESCO Assistant Director General for Natural Sciences to discuss these proposals, and in particular to investigate for ourselves whether the conditions could be fulfilled. The proposals from Armenia, Iran and Turkey were explored at the Interim Council and other meetings. Oman did not follow up. Israel had said right from the beginning that it would not propose a site because of access issues.

It quickly became clear that although the Egyptian proposal was serious, the approval process would involve multiple actors, and would inevitably be long. In Ramallah, we met with President Arafat, but despite early promise, it turned out that the Palestinian authority could not supply the resources for the required special contribution. On a special visit to Jerusalem, I learned of the balancing act that had to be performed between Israel's commitment to light sources like the ESRF and DESY, and SESAME. I learned from Jacob Ziv that if the lab was within a short driving distance of the Weizmann Institute, it would certainly be used by Israeli scientists. The Israelis favoured Jordan, if the Allenby Bridge over the Jordan River could be opened to Israeli scientists. Armenia already had a synchrotron laboratory, and offered it as a site, but Armenia was on the periphery of the SESAME region. The Iranian proposal was seriously considered but was discounted because Iran could not guarantee access to all scientists.

I vividly remember the visit to Jordan. After an official meeting with representatives of the Jordanian authorities and various universities, we did not seem to be any closer to receiving a formal offer that would fulfil the Interim Council's set of criteria. I was becoming quite desperate, since not only were we making little progress in Jordan, but there were also no other clear candidate sites emerging at that time. In my desperation I called a former student of mine, Isa Khubeis, who had come to Mainz from Jordan in 1958. He was a good student, had followed me from Mainz to Karlsruhe, and gone back to Jordan with a PhD. In the meantime, he had become a vice president of Al-Balqa Applied University. I asked him if he could help.

Fig. 11.10 Herwig met with Yasser Arafat, President of the Palestinian National Authority, at his home on 1 October 1999 to discuss SESAME (©CERN, All rights reserved)

Fig. 11.11 Herwig with his former student, Isa Khubeis, who played a key role in securing a site for SESAME in Jordan (Herwig Schopper's personal collection. ©Herwig Schopper, All rights reserved)

Isa invited Iaccarino and me to dinner in his home, along with the university's president, Khaled Toukan, and his wife. Later in the evening, another guest arrived, who turned out to be the King's cousin, Prince Ghazi. After I explained the idea of SESAME to him, along with my dilemma concerning the site, he went out on the terrace to make a phone call. When he came back, he announced that we had an audience with H. M. King Abdullah II at 11 o'clock the next day. The following morning, we were taken to the palace where the King declared that he was willing to host the laboratory in Jordan. When I pointed out to him the two key conditions, he confirmed his decision. As the meeting drew to a close, I asked whether we could have the offer in writing, and half an hour later, Price Ghazi handed me a letter addressed to the Director-General of UNESCO and signed by the King. It offered a site in Allan, close to a campus of Al-Balqa Applied University, and guaranteed to fulfil the conditions stipulated by the interim council.

The formal site decision was taken at a restricted meeting of the Interim Council on 10 and 11 April 2000 at CERN. It was a very special meeting in that a single person represented each of the Interim Council members, and each had the authority of their government to take a decision. Through a series of secret ballots, the number of sites was reduced. Egypt and Iran withdrew their proposals before the final vote. In the final secrete vote the main choice was between the Allan site in Jordan, which had

Fig. 11.12 Herwig with Prince Ghazi bin Muhammad of Jordan in 2001 (Nuovo Cim. 40, 199–239 (2017) [3] ©Springer, All rights reserved)

the King's support, and the Yerevan site in Armenia for which a wealthy Armenian expatriate living in the US, Hirair Hovnanian, had offered to cover the necessary expenses. The result was in favour of Jordan. I often wonder whether SESAME would have succeeded without the help of my former student Isa, and the promise of King Abdullah II, which has been more than fulfilled."

References

1. https://twas.org
2. https://salam.ictp.it/salam/documents/one-hundred-reasons
3. Schopper H (2017) The light of SESAME: a dream becomes reality. Riv Nuovo Cim 40:199–239
4. UNESCO. Executive Board, 162nd, 2001 [361]. https://unesdoc.unesco.org/ark:/48223/pf0000123351
5. https://www.sesame.org.jo
6. https://eur-lex.europa.eu/legal-content/EN/TXT/HTML/?uri=LEGISSUM:r15006
7. https://new.worldacademy.org
8. A facility for tumour therapy and biomedical research in South-Eastern Europe, CERN Yellow Report, 2/2019. https://cds.cern.ch/record/2688922?ln=en
9. SEE-LS: A 4th generation synchrotron light source for science and technology, CERN Yellow Report, 1/2020. https://cds.cern.ch/record/2715352?ln=en
10. La Revista del Nuovo Cimento della Società Italiana di Fisica, vol 40(4). https://doi.org/10.1393/ncr/i2017-10134-8

11. European Research Infrastructure Consortium (ERIC) https://research-and-innovation.ec.eur
    opa.eu/strategy/strategy-2020-2024/our-digital-future/european-research-infrastructures/eri
    c_en

Chapter 12
In His Own Words: Epilogue and Reflexions

You may ask me whether, when looking back, I consider my long life of 100 years to be a happy one, and am I satisfied with the outcome? Despite many up and downs, and a very difficult period during and after the Second World War, the answer is yes. I have been privileged to have had a joyful youth, a long and happy marriage, and a wonderful family (Chap. 1). Through a mixture of good fortune and good health, I have frequently found myself in a position to make decisions regarding areas of science and other fields that I have always felt passionately about. Science, of course, has always been at the centre of my life. In this final chapter, I would like to reflect on the essential points, and share some conclusions drawn from my long experience of a changing world.

Science at the Centre of My Life

The main focus of my career has been research in physics, which brought me the greatest satisfaction. The joy of realisation that, as the result of painstaking work, you are the first to know something that was not known before is indescribable. Perhaps it can only be compared to the feeling felt by an explorer who voyages for the first time into unknown territory. Of course, moments like this are rare, but they nevertheless make all the effort worthwhile.

© The Author(s) 2024
H. Schopper and J. Gillies, *Herwig Schopper*, Springer Biographies,
https://doi.org/10.1007/978-3-031-51042-7_12

The Importance of Fundamental Science and Technology in a Changing World

Since the beginning of human consciousness, our species has always asked questions: what is the structure of the heavens, how do we account for the motion of the Earth, the Sun and the Moon? More recently, we have pondered the galaxies and their creation, how the cosmos started—will it have an end, and what role do Black Holes play? Today, similar fundamental questions are also being asked about the microcosm. Are there ultimate eternal building blocks of matter? Can they be subdivided into ever smaller particles? What keeps them together and guarantees the existence of matter? There are permanent changes in nature: what causes them? The attitude, and human spirit, that lead us to pursue such questions was best expressed in my opinion by Johan Wolfgang Goethe, who called it *Faustischer Drang*, Faust's urge, to understand what keeps the world together. Faust's urge came upon me early and has stayed with me my whole life.

In my lifetime, and to my great fascination, the exploration of the cosmos and of the microcosm have become closely linked together. One cannot understand the infinitely large without understanding the infinitesimally small and vice versa. Today, the natural sciences have evolved into numerous highly specialised disciplines, but physics and mathematics remain at their core, providing a unifying foundation. This is what makes them so interesting.

It may seem that pursuing such questions of fundamental science is the province of an intellectual elite, but it has always been an essential part of human culture. What would we teach our children if there were no progress in acquiring additional and new fundamental knowledge? Would we still tell them that the universe is about 4000 years old? Or that the stability of matter and its eternal changes in nature are due to some supernatural entities rather than elementary forces in nature? Spiritual concepts, not only material achievements, are important for human identity, but they must exist alongside the learnings of science.

A second argument for the importance of fundamental science is the fact that all modern technologies are the offspring of fundamental scientific discoveries. All present-day electrical applications owe their existence to the work of scientists like Alessandro Volta, Georg Ohm, André-Marie Ampère, Hermann von Helmholtz, Heinrich Hertz, James Clerk Maxwell and many others. Modern TV, telecommunications, computing and many forms of medical diagnostics rely on quantum mechanics, which was developed about 100 years ago and is one of the most abstract of theories.

Modern technologies have thus benefitted from fundamental research but the inverse is also true: many of these technologies have also led to the development of essential tools for fundamental research. Take, for example, electron microscopes, particle accelerators in their many different forms, and telescopes. It is therefore perhaps not surprising that I took an interest in both fundamental and applied science, and my own research work was carried out in both fields. Indeed, I was never fully satisfied with abstract ideas alone; I wanted to create something of tangible value as well.

Scientific Careers in Changing Times

The way a scientific career evolves has changed dramatically over recent decades. My most important publications carry only my name. This is almost unthinkable in most fields of science today where most of the research must be carried out in collaboration with others, often involving teams from several universities. In extreme cases, such as some of the experiments performed at CERN's Large Hadron Collider (LHC), the number of authors can reach several thousand. This is a practice that I tried to resist when I was Director-General of CERN in the 1980s and the experimental collaborations preparing for the LHC's predecessor, the Large Electron-Positron collider (LEP), were coming together. I failed completely. Before LEP, large collaborations consisted of a few tens of scientists, and often took their names from the initials of the collaborating institutes—CDHS, for example: CERN–Dortmund–Heidelberg–Saclay—which was led by the Nobel Prize winner, Jack Steinberger. LEP experiments reached several hundred collaborators, and their names had evolved into sometimes-contrived acronyms describing their detectors—DEtector with Lepton, Photon and Hadron Identification, DELPHI, for example.

Other practices have vividly changed also. Today everyone from a budding scientist to someone aiming for the top jobs has to apply for any position by filling out forms, writing CVs and quoting their publication statistics. All these measures are introduced to make the selection procedure more objective and transparent, but I

Fig. 12.1 Herwig receives the Albert Einstein Gold Medal from UNESCO Director-General Kōichirō Matsuura in April 2004. The Albert Einstein Gold Medal is a high distinction that UNESCO confers on outstanding people who have made a major contribution to science and international cooperation (©UNESCO, UNESCO Photobank, CC SA 3.0 IGO)

Fig. 12.2 Herwig was later awarded UNESCO's Niels Bohr Gold Medal, which was presented to him by Danish Minister, Helge Sander, at the Royal Danish Academy of Sciences and Letters on 15 November 2006. The medal is awarded to "researchers who have made outstanding contributions to physics—research which, furthermore, has or could have a significant influence on the world" (©Royal Danish Academy of Sciences and Letters, All rights reserved)

doubt it's the best way. Anyway, I have never applied for a job in my life, so I don't believe I would be able to make a reasonable career today!.

My passion for science has helped me to overcome many difficult moments in my life and has contributed essentially to shaping my personality. I have often recalled the words of my father who told me that: "in moments of distress, work and the engagement in it can always be the harbour of refuge." Certainly, my work has given me the satisfaction of having made at least some minor contribution to human history. Of course, reputation and even glory are very ephemeral. With very few exceptions, people and their achievements disappear into archives or even oblivion. I sometimes wonder what future archaeologists would make of CERN. Maybe the LEP tunnel will be the part of my legacy that will last longest. Nevertheless, I suspect that anyone excavating CERN hundreds of years from now and finding a circular tunnel with a precisely defined geometrical shape, completely useless for any kind of traffic, would be a little bit baffled. They would probably conclude that it was a kind of place of worship, like a medieval cathedral or Stonehenge!.

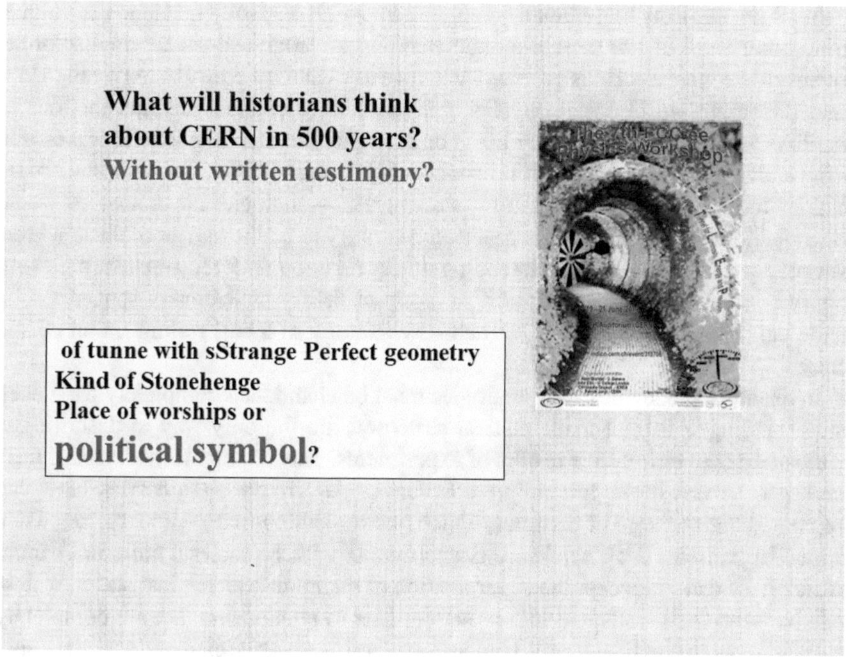

Fig. 12.3 In a slide from Herwig's talk at the LEP closing ceremony, he speculates on what future historians might make of the LEP/LHC tunnel were they to find it without accompanying documentation. Would they look at it as we look at Stonehenge? (Herwig Schopper's personal collection. ©Herwig Schopper, All rights reserved).

Public Understanding of Science

Based on my own experience, certainly limited but gathered over many years, I have developed some general thoughts that I would like to share. At the same time, however, I am very much aware how difficult it is to pass on one's own experience to later generations. Young people consider the words of their elders largely as quaint stories with little relevance to them. I certainly did, and it is only later in life that we learn to listen to our predecessors. I hope that some of my thoughts and reflections might induce at least some readers to develop further their own ideas.

It seems to me that it is relatively easy for journalists to explain to the public the beauty of the arts, the basic principles of economics, or the rules of sport. To my regret, a real understanding of fundamental physics such as quantum mechanics, and in particular of elementary particle physics and astrophysics, requires a long and challenging apprenticeship. A book containing a single mathematical equation will lose about half of its readers. This is not to say that an appreciation of the beauty of science is not possible without a grasp of mathematics. It is, and there are some wonderful communicators of science. Nevertheless, to me it is a pity that so many people miss out on the pleasure and the beauty that natural sciences can provide.

In all the subjects that humans communicate with each other, science is unique. In no other walk of life is public understanding a recognised academic discipline. However, we meet a serious problem when we ask what we really mean by the word 'understanding.' In classical physics it is possible to connect abstract theories to familiar experiences in everyday life. For example, the electromagnetic waves that bring television into our homes and connect our mobile phones can be described by analogy to water waves. The analogy is not perfect—analogies never are—as we no longer believe in the ether of the nineteenth century to be the carrier of these waves. Nevertheless, as a tool to understanding, it's good enough. Even Einstein's general theory of relativity, which requires the abstract theory of Riemann space, can be made tangible—*begreifbar* in German—by analogy to a ball rolling on an elastic sheet.

In quantum mechanics such analogies must be abandoned completely for a deep understanding. Abstract mathematical structures are the only way to describe the completely counterintuitive results of experiments. Any attempt to understand fully these phenomena through a simple concrete or descriptive picture fails. How can we understand the spin of particles, which has something to do with rotation, when point-like particles such as electrons and quarks with no internal structure cannot rotate? How can we accept the superposition of quantum–mechanical states, which, together with their statistical interpretation, lead to phenomena that are completely unimaginable in classical terms. Human logic and experience cannot do it—it is only mathematics that can make sense of what we observe, and mathematics is largely immune to analogy.

This total counterintuitiveness was one of the reasons why Einstein, and many other physicists, never accepted quantum mechanics as an ultimate theory. If it challenged Einstein, then it's no surprise that quantum mechanics is so impenetrable to most people. Are fundamental scientists, theoretical and experimental, becoming modern-day monks, respected by the outside world but isolated in our spiritual monasteries? I was once struck by the comments of a member of the audience for a public lecture I gave about the Higgs particle, CERN's most recent great discovery. After I'd finished, she said to me: "Professor, your talk was excellent although I did not understand much. However, my non-understanding was at a much higher level than ever before." So, on the one hand, these abstract ideas can be considered as the pinnacle of human thought, imagination and intellectual achievement. On the other hand, what is that worth if it is inaccessible to most of humanity? The problem is that these developments are based on abstract concepts and most people abhor abstract thinking. Nevertheless, sometimes the most abstruse ideas can prompt imagination in cultural works. Recently a film based on the concept of parallel worlds in quantum mechanics was an Oscar-winning box office hit.

Of course, I'm playing devil's advocate here. As I've already said, much of the technology we take for granted in modern life relies on abstract physics such as this. I could equally well argue that it doesn't matter whether people fully understand it or not. But I think it does matter. We scientists must never give up on our efforts to explain and to engage with the public, and to share with them the importance of science as part of our culture.

Another, equally important, aspect of public understanding of science is how science works. Natural sciences like physics, chemistry and biology play a steadily increasing role in society, underpinning many modern technologies that have changed our daily lives. As a society, we take this for granted, yet unfortunately, the general public does not understand how science works. People turn to science for accurate predictions of what will happen in the future, but science cannot do that. Science can make firm statements based on accumulated evidence from experiments that must have been shown to be reproducible, but it cannot make predictions with the same degree of certainty.

Sensationalism and Scientific Revolutions

Research is a never-ending endeavour, with each new result raising new questions. Scientific revolutions, as media like to call them, do not really exist in the sense that everything previously known becomes obsolete. Rather, these so-called revolutions usually imply that the laws which are valid in a certain domain are limited to this domain, while different laws are valid in another. For example, quantum mechanics has revolutionised science, but it has not invalidated the classical physics that went before. When we are dealing with the kind of scales or dimensions of objects that we are familiar with from daily life, metres and centimetres, classical physics works just fine. If we deal with single atoms or molecules that are more than a million times smaller, then we must apply quantum mechanics. This is the nature of a scientific revolution: it just tells us that different laws have to be applied in different domains. They do not contradict themselves, but one set of rules is the special case of more general laws. We say that quantum mechanics yields asymptotically classical physics when we go from atomic dimensions to everyday life. It would be a disaster if classical physics would be devaluated completely by quantum mechanics. Radio, television and mobile phones would not work, and aircraft would not take off anymore. So when the media announce scientific revolutions, there's no need to be shocked!

To give another example, classical physics is valid when dealing with velocities that are much smaller than that of light. If we look at objects moving with velocities close to the velocity of light, then we must use the special theory of relativity. That is the case at CERN, where the accelerated particles travel with velocities very close to that of light. This leads to many misunderstandings.

For example, it would be impossible to accelerate a spaceship to the speed of light. According to the laws of special relativity, the mass of the spaceship would increase with its velocity and become infinite at light speed. To bring an object with any mass to such a velocity would require infinite energy. Only light, which is massless, or other massless objects, can reach that velocity. Special relativity states that the velocity of light is the limiting speed at which signals can be transmitted, using light or any other means. Nothing can travel faster.

This has implications for the question of whether intelligent life exists beyond our local cosmic neighbourhood. It's a question that does not interest me very much,

Fig. 12.4 World-famous cellist Yo-Yo Ma (left) with CERN Director-General Fabiola Gianotti and Herwig Schopper at an event exploring common aspects of music and physics at CERN in December 2023 (Courtesy Fabiola Gianotti, ©Fabiola Gianotti, All rights reserved)

because even if intelligent life does exist in other galaxies, it would take hundreds or thousands of years to get an answer. Who has that much patience? Certainly not me! Does this kind of misrepresentation of science matter? Possibly not, but isn't the science of events on our planet more interesting?

People seem to like sensation, and the media we consume pander to that, sometimes for good reasons, sometimes not. One revealing example is the first picture of a Black Hole, which made the front pages of many newspapers around the world. People were fascinated by it, without really knowing why. When young people were asked for their thoughts about it, some said that it looks like a hot doughnut, which I suppose the picture does. What most people didn't realise was the image they were looking at has little to do with any normal photograph but is based on a complicated phenomenon that is hard to explain. It is already difficult enough to explain a Black Hole, because it is a special mathematical solution of general relativity. Indeed, there are different types of Black Hole, all with the common property that they accumulate a sufficiently large mass that their gravitational attraction prevents even light or any other radiation from escaping. This means that they cannot be seen in the conventional sense of the word, and their existence can only be detected indirectly. It is a great triumph of astrophysics that their properties and their role in the universe can be deciphered at all, and this so-called photograph is just one more way of getting indirect information. As we can't see Black Holes, we can't photograph them, so this famous image is not a photograph in the classical sense, but rather a visualised representation of the data. In this case, the media succeeded in generating excitement about science, although they hardly explained the nature and roles of Black Holes, even though an extremely heavy one sits at the centre of our galaxy.

It's not always like that, however. When CERN first circulated beams in the LHC in 2008, there was enormous media interest. Over 350 media outlets were at CERN that day. Why? Some were there to cover the start of a new era of research in particle physics, but most were there for another reason. A somewhat crazy and completely unfounded speculation was doing the rounds that the LHC would produce a certain type of mini-Black Hole that would devour the world. Needless to say, this did not happen—sophisticated though it is, the LHC does nothing more than recreate naturally occurring phenomena in the laboratory where they can be studied. There was never any risk, but the sensationalism succeeded in scaring many people around the world. Without wishing to detract from the many good journalists reporting on science in a balanced and ethical way, if this is indicative of the quality of science coverage we can expect, we face the danger that science will become a kind of modern superstition—mysterious, somewhat attractive, but at the same time creating fear. Schools have a great responsibility in ensuring that this does not happen.

We are in daily contact with technologies such as television, mobile telephony, computing and networking. Like most people I am no longer able to repair them. But in principle I understand how the technology works, and I do not have to believe in ghosts or any other form of supernatural phenomena. It would be a good thing in my opinion if everyone could have a similar level of understanding of the things they take for granted, especially young people.

There's also a more important reason why it's important to understand science, and the way it works. Today, more than ever, an ability to make decisions based on the evidence available, and to change those decisions as the evidence evolves, is increasingly important for everyone. You need look no further than the recent COVID-19 pandemic for evidence of that, or the kind of decisions that society as a whole is taking when confronted with overwhelming evidence that we are having a devastating impact on the Earth's climate.

Science, Politics and the Role of Forecasts

The fact that society, and above all politicians, expect science to make definite forecasts is a problem that can have very serious consequences. Of course, science can make solid predictions based on past experience, but forecasts have their limits. This is what politics either fails to understand, or wilfully ignores: science is often used as a shield to cover political intentions. Why does this happen? I can only scratch the surface here—each of the points I'm about to raise would merit a book to itself.

To make forecasts, we have to develop models that require a number of assumptions on processes that are not yet known, or that are known with some degree of uncertainty. The obvious, and most pressing, example is climate change. Most of the processes that determine the atmospheric and oceanic climate are well known, but not all. For example, clouds influence how much sunlight is reflected back into space, while mechanisms governing the formation of clouds are extremely complex, and not fully understood. A unique experiment at CERN, which probably could not be done anywhere else, painstakingly simulates atmospheric processes to provide input to climate models and thereby refine their predictions. The CLOUD experiment—an acronym derived from the phrase cosmics leaving outdoor droplets—can simulate the atmosphere to extraordinarily high precision, even going so far as to include cosmic radiation, which is simulated by a particle beam from a CERN accelerator. This allows the experiment to investigate the mechanisms by which clouds form, and the extent to which anthropogenic activity influences those mechanisms. CLOUD has been running since 2009, and its work is far from done.

Another major source of uncertainty in climate modelling is the interaction between the atmosphere and the ocean, which absorbs a large fraction of the CO_2 in the atmosphere. These and other poorly understood processes influence the precision of any forecast, which is why forecasts for future temperature rises are quoted with an element of uncertainty. General trends can be predicted, but as to the quantitative statements beloved of politicians, we should be very sceptical. Unfortunately, politicians all too often misinterpret scientific uncertainty as evidence that scientists do not really know what they are doing, and say that what science puts forward as evidence is just an opinion, equivalent to that of those who would deny climate change. This is irresponsible, unacceptable and dangerous.

There is another inevitable uncertainty in scientific statements. Physics students learn in their first semester that a quantitative statement in science has always to be accompanied by a statement of its errors, otherwise it has little significance. One ubiquitous source of uncertainty is the so-called statistical error. In any probabilistic process, statistical error is important. Consider throwing a die. If the die is perfectly fabricated, the probability that it will land on any of its six numbers will be the same, namely one sixth, but it will take many throws to verify this. Every gambler knows, and dreams of, the same number coming up several times in a row. That does happen, but only very rarely. The laws of statistics tell us how likely the average result of any measurements is, and how likely it is that we will see deviations from this average. The precision of a measurement increases as one over the square root of the number of times it is repeated. Hence, with 100 measurements the statistical error would be 10% and for 1000 trials it is about 3%. A clarification of the terminology may be useful here. What we call 'error' in measurements is not associated with any wrongdoing as in daily life, rather, it signifies a level of uncertainty.

The Gaussian distribution is one of the most used formulae in statistics, and it even appeared on a 10 Deutschmark banknote in the 1990s. The Gaussian gives us information on how probable deviations from the most likely average value are. Extremely rare events in the so-called tails of the probability distribution are never-theless possible. These are the events that some decision makers, and certain elements of the media, like. As no scientist can say that they are impossible, such extremely unlikely events allow agendas to be pushed and sensations to be made.

Statistics is not the only source of error. In addition, there may be other external influences. A die might be imperfectly manufactured, for example, with a tendency to favour one number more than the others. To account for factors such as this, a systematic error must be added, and its size must be estimated, which is sometimes rather difficult. Bear this in mind the next time you read a newspaper article or listen to a politician quoting a scientific number to support an opinion or justify a political decision. Ask yourself, was the number they hold up as evidence accompanied by two errors, statistical and the systematic? I suspect that the answer will be no.

This is normally true in the reporting of public opinion polls. By the time the results are reported, few of us get to see the details of the analysis, or even the margin of error on the numbers quoted. For those of us versed in statistics, we can get a feel for what the statistical error might be from the number of people polled, and of the systematic error from the demographic spread of the interviewees. But most people are not trained in statistics, so are ill equipped to interpret the numbers they are given. Because of limitations in time and resources, even a large public opinion poll probably will not interview more than a few thousand people. If the poll is about voting intentions and the results for two parties differ by under a few per cent, then no conclusion can be drawn about voting intentions, even before taking systematic errors into account. Of course, the experts who conduct the surveys know this, but by the time the results make it onto the TV news, they are simplistically reported as party X leads party Y by a few per cent in the polls. In other words, differences that have absolutely no significance are discussed in the public sphere at length, and this can influence the outcome of the real election.

Turning back to climate change, the neglect of errors has very serious conse-quences. Anyone who follows the news at all is familiar with the debate around whether the permissible rise in global temperature to avoid catastrophe is 2° or 1.5°. It's the wrong question. In the public debate, I have never heard about the scientific uncertainty on these figures, but if you read the original publications of course, you find that the errors are several degrees [1]. In the light of this, it makes no sense to argue about a difference of 0.5°, but that's what the world is doing.

What we should be focusing on is the overwhelmingly strong scientific evidence that the climate is warming rapidly, and that we have a lot to do with that. Whether it is by 1.5° or 2° is irrelevant given that the uncertainty on individual forecasts is of the order of several degrees.

The difficulty in making detailed predictions, and the potential pitfalls in trying to do so, became painfully evident when the Club of Rome rightly made the forecast that there are limits to growth. They predicted that it would not be possible for the agricultural production of the planet to feed 10 billion people, and that at a time when there were no climatic problems in general sight. This shows how difficult it is to make long term forecasts with reliability. Today it seems clear that food is not the crucial limit, but rather that other restrictions are more important, such as the rapid growth of populations. This was explained clearly by Ernst Ullrich von Weizsäcker, a scientist, politician and honorary president of the Club of Rome in his remarkable keynote address at the 50th anniversary of the European Physical Society in Geneva on 28 September 2018. In a paper entitled 'Come on [2]', he explains that one has to give up the belief that current trends are sustainable and adopt new and exciting journey with long term visions. For that purpose, a kind of new enlightenment might be necessary. The first enlightenment helped to liberate our spirits and prepare them for the technical revolutions, leading to a materialistic philosophy and in the end to selfishness and brutal competition. Enlightenment 2.0 should instead concentrate on a balance between contradicting elements like human versus nature, public versus state, fast change versus stability, feminine and masculine, religion and state and others. It seems that Asian civilisation can more easily accept such balances whereas the tendency of the West is to polarise.

The Club of Rome was absolutely right to draw attention to the limits to growth, but society's reaction was to debate the details of the premise rather than address the problem. With climate change, we're seeing the same thing happening: we should be concentrating on mitigation instead of arguing about tenths of degrees. This could allow us to develop the necessary technologies instead of taking precipitous decisions with little impact. We could then set the most favourable speed to fight climate change without damaging social life worldwide. Recently it seems that it is becoming more and more evident that the 1.5° target cannot be met and more importance should be given to mitigating measures, in particular concerning rising sea levels.

In many walks of modern life, those who should be taking a lead frequently take refuge behind what they refer to as objective scientific facts. In doing so, they are shifting the responsibility they should assume on to others: the scientists who have supposedly provided those objective facts. Even worse, they establish so-called expert committees with scientific-sounding names that often appear impartial but are frequently far from it. When something goes wrong nobody is to blame but the science, which has been misrepresented by design.

In some cases, ethical or truly political issues with a scientific dimension are transferred to legal courts or parliamentary bodies that again turn to science for certainty. This kind of practice is based on the same misconception of science that I've already discussed. Science itself cannot give firm advice on decisions for the application of technologies. It can perhaps offer different scenarios for future developments. Unfortunately, some individual scientists also overestimate their own work, and play into the hands of a system looking to avoid responsibility. The solution? There is probably no simple answer, but when looking for impartial scientific evidence, perhaps the best way is to turn to national scientific academies such as the UK's Royal Society or Germany's Leopoldina, but there may be other possibilities.

Is There a Universal Truth?

Another issue that has intrigued me all my life is the notion of universal values. In science we agree on ways to establish whether a result is true or false: the final verdict on whether a theory is right or wrong is provided by reproducible experiments. Could it be possible to find a similar procedure for ethical values that would be true everywhere, for everyone and at all times?

I touched on this question in a previous chapter (see Chap. 11) when discussing the relationship between natural science, religion and art. It may suffice here to repeat that what science and religion consider to be true is based on completely different concepts. In science a true statement can always be verified by experiments at any time and in any place. Only if we explore nature under these conditions can we find rules independent of tradition, political systems and ethical considerations. Religion, on the other hand, relies on revelation. The aesthetic in the arts is yet another completely different realm of human life that involves subjective personal taste. They are all different experiences of reality, completely independent and complementary, and our lives would be incomplete and poorer without them.

Could it be possible to agree on conditions, as we have done for science, to establish universal ethical principles? I doubt it. The testing experiment would be history, which cannot simply be repeated. Hence, I believe there is no absolute truth in this sense. As far as ethical principles are concerned, everybody has the right to choose the basic elements according to their tradition, education and power of judgement. Time changes what any given society considers to be ethically acceptable. Of course, human rights are recognised today as guiding principles. But the historical development of these principles was achieved mainly in the western world, starting with the French revolution, and evolving through the social improvements during the more recent technical revolutions. We should be proud of these achievements, but we should also recognise that western cultures represent only a minority of humanity, and we should be wary of arrogance when trying to impose western value systems wholesale on other cultures.

In the west, we believe that democracy is the best way to organise society. However, in discussion with colleagues from other cultures, I have often been challenged to say what I mean by democracy. Is an elected parliament alone a guarantee of democracy? Probably not. And does democracy have to be practised as it is in the UK, France, the USA or indeed in any other western democratic country? There are essential differences in political practice, which lead to a spectrum of differences in the behaviour of societies. Finding the right balance between the rights of the individual and the interests of society is not so easy. Too much liberalism favours extreme egotism, whereas a too strong state risks stifling private initiatives and freedoms, with government rules overseeing what goes on even in our bedrooms. More tolerance is

needed in all kinds of discussions between nations and cultures, where mutual respect
and understanding are paramount. Could these be principles one might be able to
accept globally? I doubt it. Perhaps different solutions for different areas of the globe
are what we should seek: good relations with mutual respect and tolerance would
be a better solution. Nation states, a result of the French revolution and adapted to
benefit from the emerging new technologies may indeed be a bad option for coping
with the advantages and dangers of present-day technologies.

Is Technical and Social Progress Too Quick for the Human Mind?

I once had the chance to discuss politics with Werner Heisenberg. He pointed out
that in classical physics we learn that a sudden transition, a step function as we call
it, imposed on a system in equilibrium leads to violent oscillations that die away only
after some time, which is determined by internal parameters of the system. If you
want to move a system smoothly from one steady state to another you have to do it
slowly, adiabatically as we call it in physics, in order to avoid turbulence. The same
may also be true for social systems: if we talk about the evolution of events in time,
it's important to be aware that developments need a certain natural time to happen
smoothly.

I think that turbulence can happen in social systems if traditions and mentalities
are changed too rapidly. Revolutions lead to turbulence and violence, and as the
old adage tells us, they devour their own children. Humans, it seems, as well as
natural phenomena, need time to adapt to new conditions. For that reason, it is
rarely productive to impose a new ethical value system on a society with a particular
mentality and traditions, even if convinced that doing so will be beneficial for that
society. Each system has its own natural inertia, which should not be neglected.

In my travels over many decades, I have been able to observe this for myself.
When I visited Turkey 50 years ago, for example, the large towns all had western
features. The radical reforms of Mustafa Kemal Atatürk had swept away centuries of
Ottoman tradition and replaced it with westernisation. Men were no longer obliged to
wear long beards, writing switched to the western alphabet, and the political system
was democratic. Every time I have visited Turkey since, I have found that it has taken
some steps back towards the old value system. Atatürk introduced turbulence, and
it will still take time for the new Turkey, or Türkiye as the country is now known at
the UN, to settle into a new equilibrium. I have observed similar developments in
some countries in the Middle East. In Russia, Communism did not extinguish the
old orthodox religion. And in Afghanistan, decades of intervention failed to supplant
the tribal mindset that prevails again today.

Even if elements of a system are distasteful to our value system they cannot be
changed by force. I remember a dispute with a Chinese minister at a Forum Engelberg
meeting, where someone criticised the corruption in his country, He replied: what

is corruption? What you call corruption is part of our tradition for 2000 years. He warned us that too much pressure on changing their system could be considered as neo-imperialism. One should not lightly condemn other social systems, before fully understanding their history, and arrogance should be avoided.

The influence on society of the rapid advance of technology may be less obvious than that of revolutions, but it is nevertheless there. It has rapidly changed the mentality of large parts of humanity, like mobile phones in developing countries to cite one example.

What Are the Priorities in Politics?

I have never been a member of a political party or carried out a political job. I have, however, been involved with politicians throughout my career, and this has allowed me a glimpse into their mentality. It has often surprised me how fast they sometimes have to change their priorities, and I'd like to recount two particularly striking examples.

When I was at DESY, I got to know the publisher of the weekly newspaper *Die Zeit*, Marion Gräfin von Dönhoff, who was one of the most influential journalists in Germany, and Helmut Schmidt, who had visited DESY when he was Federal Chancellor of Germany and joined *Die Zeit* as co-publisher later. Dönhoff had the intention of starting some supporting activities for large projects, and I became involved in her circles. As a result, I was invited to a private meeting at of about a dozen people at a hotel in Berlin on the occasion of her 75th birthday in 1984. The star guests were none other than Helmut Schmidt and Henry Kissinger. They had both retired from politics and they gave talks about what should have been done to improve the global political situation. When we asked why they did not follow their own advice when in office, the answer was that when one takes up an executive position the priorities change: the main objective is to stay in power and win the next election.

This struck me very deeply, but it means that I was not too surprised when I got a similar answer a few years ago when I was invited to a meeting in Baku, Azerbaijan. The president had invited about two dozen former presidents or prime ministers from east or south European countries to a meeting where only they were allowed to speak. A few scientists were also invited but we could only listen during the official sessions. I had a sense of déjà vu listening to them explaining with great conviction what should be done to improve the global political situation. When I asked them during the coffee breaks why they did not do these things when in power, I got the same answer as before: priorities change drastically when one is appointed to an executive job. As a party leader in opposition, one can propose many ideas without being compelled to prove immediately that they would work in practice. Promises during an election campaign are something different from establishing concrete laws with immediate consequences. From these discussions I learned that one cannot expect the same objectivity and sustainability that one is accustomed to in science.

A World in Transition

A visiting alien looking at the situation of the world today and listening to some self-appointed prophet of doom would be forgiven for concluding that the end is nigh. I tend to disagree. Having lived through, and experienced at first hand, the horrors of global conflict, I'm actually more optimistic about the future than you might expect.

It is true that society faces multiple challenges that appear existential, and the majority of humanity has living conditions far below those that the industrialised countries enjoy. Modern information technologies make this strikingly clear to anyone who cares to look, and it's not unreasonable for those in poverty to aspire to western standards of living. Added to this, climate change is having a drastic impact on the planet, leading to extreme weather events happening with increasing frequency, and even changing the regions of the planet that are hospitable to human habitation. If no satisfactory solution can be found, the migration that we see today will soon seem like a trickle. Turning migrants around at our borders is not the solution. We need to think more rationally and inclusively about dealing with such challenges.

Claims that the end is nigh are nothing new. They are as old as human civilization itself, enshrined in religious texts the world over. The Mayan calendar, for example, famously ended around 2012. Today, it is the *Bulletin of the Atomic Scientists* that keeps us on our toes, updating their famous doomsday clock regularly to show how close we are to midnight—the moment it all ends. Over the three-quarters of a century of its existence, the clock's time has moved closer or further away, but it has always been very close to midnight. At the time of writing, we are just 90 seconds away. I don't suppose that the scientists behind the clock truly believe that the world will end at their symbolic midnight. Rather, they are keeping the issues that we need to be dealing with in the public eye, exploiting our love for the sensational that I've already discussed. In a certain way, this is a good thing: it helps us to focus on the need for solutions, but in heeding their warnings, we must also remember that rushed decisions will not solve our problems. We need well thought-out solutions.

History teaches us that there are two ways to change existing systems. There is the abrupt cataclysmic approach exemplified by revolutions, or there is peaceful evolution that takes time. In the latter we can adapt in small steps, taking sufficient time for society and nature to evolve in parallel. This implies giving sufficient time to develop new technologies and learn how to live with them. Such advances cannot simply be ordered, as the Soviets tried to do with their 5-year plans, for example.

Time is short, but we are an innovative species. I for one am inspired by the energy, ideas and creativity of generations much younger than my own, and that makes me optimistic. We are living through a time of transition and those younger generations give me cause for hope.

Whichever way the world changes, peacefully or not, no country or region can go it alone. Natural systems have always been interconnected, and modern technology means that so are we. No one is immune from the actions of others, and it is in everyone's interest to work together. One thing is certain, an element of competition between countries and regions will always be present, and major players on the world stage, such as the USA, China, India, Africa and South America—and I hope also Europe, will have major parts to play. They will need to work together to ensure sustainable peaceful coexistence, not only relying on formal rights, but also on balance based on mutual understanding and tolerance. This requires a positive narrative and optimism instead of belligerent competition set against a background of doomsday scenarios.

I have the impression that today's pessimism is mainly present in the developed countries. In particular, the younger generations are frustrated, seeing little perspective for a better future. The only long-term prospect they see is climate change and its attendant problems. In countries where social improvements are still possible, the mind-set is completely different. I have seen this in my travels to developing countries, and to countries with large populations such as India and China. In such places, the younger generations are motivated to work hard and develop plans for a brighter future for themselves, or at least for their children. Such optimistic narratives are missing for the developed world.

What can be done? Some people claim that scientific and technological developments will not be crucial for social progress. The history of the past few hundred years suggests otherwise. I do not say that such progress alone will suffice, but without it, it will be very difficult to avoid a major human disaster.

Our leaders would do well to look at global science for inspiration. As anyone working in science knows, scientific fields thrive through a blend of collaboration and competition, hard work and great effort. Science is international, and without discrimination of traditions, religions or races. Even if political solutions will be delimited by region, the whole world will benefit from the results of global science and technology.

In my life and career, I've been lucky to count as friends people from many cultures. Whatever their religious, cultural or political background, a shared vision of progress for humanity has united us. These are lofty words, and whatever happens, the Earth will survive. Future generations will find themselves confronted by their own challenges, and as all generations have done, they will find their own solutions.

References

1. Schopper H (2016) Scientific knowledge and the citizen, 26 Oct 2016. https://cadmusjournal.
   org/article/volume-3/issue-1/scientific-knowledge-and-citizen
2. von Weizsäcker EU, Wijkman A (2018) Come on! Springer, 978-1-4939-7418-4. https://link.
   springer.com/book/10.1007/978-1-4939-7419-1

Some Important Publications by Herwig Schopper

The Inspire HEP database lists several hundred publications bearing Herwig's signature. This list is a selection of those that Herwig considers to be his most important.

- **Optics and condensed matter**

  Fleischmann R, Schopper H (1951) Die Bestimmung der optischen Konstanten und der Schichtdicke absorbierender Schichten mit Hilfe der Messung der absoluten Phasenänderung. Z Phys 129:285. https://doi.org/10.1007/BF0132 7483.

  First method to measure the absolute phase on metal layers.

  Schopper H (1953) Zur Deutung der optischen Konstanten der Alkalimetalle. Z Phys 135:163. https://doi.org/10.1007/BF01333340.

  The anomalous optical behaviour of thin alkali metal layers does not require a new phase of the metal (as claimed by the Pohl school).

  Schopper H (1955) Die optische Untersuchung der Diffusion von Metallen ineinander. Z Phys 143:93. https://doi.org/10.1007/BF01330697.

  A new method to study the diffusion of metals in very thin layers.
- **Nuclear physics**

  Clausnitzer G, Fleischmann R, Schopper H (1956) Erzeugung eines Wasserstoffatomstrahles mit gleichgerichteten Kernspins. Z Phys 144:336. https://doi.org/10.1007/BF01340806.

  First prototype of polarised ion source.

  Schopper H (1957) Circular polarization of γ-rays: further proof for parity failure in β-decay. Phil Mag 2:710. https://doi.org/10.1080/14786435708242717.

H. Schopper and J. Gillies, *Herwig Schopper*, Springer Biographies, https://doi.org/10.1007/978-3-031-51042-7

One of the three experiments (thought to be unfeasible) proposed by Lee and Yang to test parity violation. The first experiment to show that helicities of neutrino and antineutrino are opposite.

Schopper H, Galster S (1958) The circular polarization of internal and external bremsstrahlung. Nucl Phys 6:125. https://doi.org/10.1016/0029-5582(58)900 84-1.

First experiment to measure circular polarisation of internal bremsstrahlung in β-decay.

Schopper H, Müller H (1959) Lepton conservation and time reversal in beta-decay. Nuovo Cim 13:1026. https://doi.org/10.1007/BF02724830.

Among the first phenomenological papers to investigate time reversal invariance.

- **Accelerator technology**

Halbritter J, Hietschold R, Kneisel P, Schopper H (1968) Coupling losses and the measurement of Q-values of superconducting cavities, KFZ Karlsruhe, KFK-Bericht 758.

One of first papers to study superconducting cavities for accelerators.

Schopper H (1969) Optimization of superconducting RF particle separators. In: Proceedings of the 7th international conference on high-energy accelerators (HEACC 69), Yerevan, USSR, 27 Aug–2 Sept 1969, pp 662–668. https://inspir ehep.net/literature/60117.

New ways for particle separation and acceleration.

- **Elementary particle physics**

Littauer RM, Schopper H, Wilson RR (1961) Structure of the proton and neutron. Phys Rev Lett 7:141. https://doi.org/10.1103/PhysRevLett.7.141; 7:144. https://doi.org/10.1103/PhysRevLett.7.144.

It was shown that the original analysis of the measurements of nuclear formfactors by Hofstadter et al. were wrong.

Behrend HJ, Brasse FW, Engler J, Hultschig H, Galster S, Hartwig G, Schopper H, Ganssauge E (1967) Elastic electron-proton scattering at momentum transfers up to 110 fermi^{-2}. Nuovo Cim A 48:140. https://doi.org/10.1007/BF02721349.

First measurements of nucleon formfactors to high momentum transfers (first measurements at DESY).

Engler J, Flauger W, Gibbard B, Mönnig F, Runge K, Schopper H (1973) A total absorption spectrometer for energy measurements of high-energy particles. Nucl Instr Meth 106:189. https://doi.org/10.1016/0029-554X(73)90063-3.

Invention of hadron calorimeter (review of earlier measurements).

Böhmer V, Engler J, Flauger W, Keim H, Mönnig F, Pack K, Schopper H, Babaev A, Brachmann E, Eliseev G, Ermilov A, Galaktionov Y, Gorodkov Y, Kamishkov Y, Leikin E, Lubimov V, Shevchenko V, Zeldovich O (1975) Neutron-proton elastic scattering from 10 to 70 GeV/c. Nucl Phys B 91:266. https://doi.org/10.1016/0550-3213(75)90470-8.

The first measurements of neutron-proton scattering at high energies (Serpukhov and CERN ISR).

L3 Collaboration (1992) Inclusive J production in Z^0 decays. Phys Lett B 288:412. https://doi.org/10.1016/0370-2693(92)91121-O.

Main author of this LEP L3 publication.

- **General publications**

Schopper H (2022) Science for peace? More than ever! CERN Courier 62(5):49. https://cds.cern.ch/record/2826497.

Schopper H (2016) Scientific knowledge and the citizen. Cadmus 3(1):84. http://cadmusjournal.org/node/585.

- **Other assorted articles and presentations**

Schopper H (1998) Die Manager sind die wahren Könige (I). Politik Wirtschaft 20(8):31.

Schopper H (1989) Keine menschenwürdige Zukunft ohne Technik. Bull Schweiz Elektrotech Ver 80:1379. https://doi.org/10.5169/seals-903734.

Schopper H (1990) Wir bestehen letztlich aus Symmetrien. Tages Anz 11.

Schopper H (1991) Wie zwei Beeren sich in einen Fisch verwandeln. Die Welt 1.

Schopper H (1991) Zerbrochene Symmetrien halten die Welt zusammen (I). Basler Ztg 25.

Schopper H (2003) Thatcher and me. Phys World 16(3):19–20. https://doi.org/10.1088/2058-7058/16/3/27.

Schopper H (1991) An den Grenzen unseres Weltbilds, Gilt unsere Physik überall im Universum? Bild Wiss 28(5):60.

Schopper H (1991) Abschied vom materialistischen Weltbild: die Physik hat einen transzendenten Hintergrund. Die Welt 125:1.

Schopper H (1999) Lebenszeiten im Mikrokosmos—von ultrakurzen bis zu unendlichen und oszillierenden, Plenarvortrag, Jahresversammlung, Deutsche Akadmie der Naturforscher Leopoldina, Halle (Saale), 26/29 März 1999. Nova Acta Leopold NF 81(314):109.

Schopper H (1998) Qu'est la verité en science et en religion? In: Sciences et valeurs: ombres et lumières de la science au XXIe siècle: international conference, University of Verona, May 1998 (Organized by UNESCO and l'Association Descartes, Edition EDK, Paris 1999).

The fundamental building blocks of matter and CERN, Lecture at the gymnasium at Lanskroun, 28 May 2010.

From scientific dialogue to international cooperation Keynote speech, Forum Science for development, 9–10 May 2018, qui Nhon Vietnam, and intervention at meeting with the President of Vietnam om 11 May 2018 at Hanoi.

- **Books (author and editor)**

Schopper H (1962) Das Protonen-Synchrotron mit schwacher Fokussierung. In: Schopper H, Kollath R (eds) Anwendung der Teilchenbeschleuniger. Teilchenbeschleuniger, 2nd edn., Vieweg & Sohn, Braunschweig, pp 162–184, 314–323. Translated Edition (1967) Particle accelerators. Pittman Press, London.

Schopper H (1966) Weak interactions and nuclear beta decay. North-Holland, Amsterdam, 417p.

Citron A, Schopper H (1970) Superconducting proton linear accelerators and particle separators. In: Lapostelle PM, Septier AL (eds) Linear accelerators. North-Holland, Amsterdam, pp 1141–1180.

Schopper H (ed) (1973) Elastic and charge exchange scattering of elementary particles. Landolt-Börnstein New Ser Group I 7. https://doi.org/10.1007/b19943.

Schopper H (ed) (1975) The investigation of nuclear structure by scattering processes at high energies. In: Proceedings of the international school of nuclear physics, Erice, Sicily, Italy, 22 Sept–1 Oct 1974. North Holland, Amsterdam.

Schopper H (1985) CERN and high energy physics. AIP Conf Proc 134:74. https://doi.org/10.1063/1.35440.

Schopper H (ed) (1993) Advances of accelerator physics and technologies. World Scientific, Singapore. https://doi.org/10.1142/1650.

Schopper H (1996) Experimente zum Standard-Modell der Teilchenphysik in Didaktik der Physik. DPG Fachverband Didaktik, Jena.

Jacob M, Schopper H (eds) (1995) Large facilities in physics. In: 5th EPS international conference, Dorigny, Switzerland, 12–14 Sept 1994. World Scientific, Singapore. https://doi.org/10.1142/9789814533126.

Schopper H (2008) SESAME—a project to foster science and peace and its relevance for the region. In: Aslam J, Hussain F, Riazuddin (eds) Contemporary physics 2007, proceeding of the international symposium, Islamabad, Pakistan, 26–30 Mar 2007. World Scientific, Singapore, pp 21–33. https://doi.org/10.1142/9789812818942_0002.

Schopper H (2009) LEP—the Lord of the collider rings at CERN 1980–2000. Springer, Berlin. https://doi.org/10.1007/978-3-540-89301-1.

Schopper H (2011) Elementarteilchen—oder woraus bestehen wir? In: Martienssen W, Röss D (eds) Physik im 21. Jahrhundert. Springer, Berlin, pp 325–365. https://doi.org/10.1007/978-3-642-05191-3_9.

Schopper H (ed) (2012) CERN's accelerators, experiments and international integration 1959–2009. EPJ H 36(4):437–632. https://epjh.epj.org/component/toc/?task=topic&id=88.

Schopper H, Di Lella L (eds) (2015) 60 years of CERN experiments and discoveries. World Scientific, Singapore. https://doi.org/10.1142/9441.

Editor of Landolt-Börnstein Tabels and Functions Kernphysik und Kerntechnik from 1967 vol 2, from 1991 vols 13–26 Kern- und Teilchenphysik. Springer.

Index

© The Editor(s) (if applicable) and The Author(s) 2024
H. Schopper and J. Gillies, *Herwig Schopper*, Springer Biographies,
https://doi.org/10.1007/978-3-031-51042-7

Printed in the United States
by Baker & Taylor Publisher Services